T5-AGQ-773

Ethics of New Reproductive Technologies

Also by Jonathan Glover

Responsibility
Causing Death and Saving Lives
What Sort of People Should There Be?
I, The Philosophy and Psychology of Personal Identity
Philosophy of Mind (ed.)

STUDIES IN BIOMEDICAL POLICY

Ethics of New Reproductive Technologies

The Glover Report to the European Commission

Jonathan Glover
& others

Northern Illinois University Press
DeKalb 1989

Published by the Northern Illinois University Press, DeKalb, Illinois 60115

ISBN 0–87580–147–1

Simultaneously published in England under the title *Fertility and the Family*.

Library of Congress Cataloguing-in-Publication Data
Glover, Jonathan.
Ethics of new reproductive technologies.
(Studies in biomedical policy)
1. Human reproductive technology. 2. Human reproductive technology – Moral and ethical aspects.
I. Title. II. Series. [DNLM: 1. Embryo Transfer.
2. Ethics, Medical. 3. Family. 4. Fertilization in Vitro. 5. Genetic Intervention. 6. Mothers.
WQ 205 G566F]
RG133.5.G57 1989 362.1′98 88–34523
ISBN 0–87580–147–1

Manufactured in Great Britain by Billings & Son Ltd, Worcester

Contents

Members of the Working Party

Professor Heleen Dupuis,
Faculty of Medicine, University of Leiden.

L.R. Karhausen, M.D.,
Commission of the European Communities.

Dr. Simone Novaes,
Centre de Sociologie de l'Éthique, CNRS-IRESCO, Paris.

Professor Povl Riis, M.D.,
University of Copenhagen. Physician-in-Chief, Medical Gastro-enterological Department C, Herlev University Hospital.

Dr. med. Bettina Schöne-Seifert, M.A.,
Institut für Geschichte der Medizin, Georg August Universität, Göttingen.

Dr. Labrini Veloyanni-Moutsopoulou,
Lawyer and Ethicist, Ioannina.

The Working Party was chaired by
Jonathan Glover,
New College, Oxford.

The Working Party wishes to express its gratitude to Dr. Robert Winston, of the Hammersmith Hospital, London, for his help in providing information.
Dr. Schöne-Seifert wishes to express her gratitude to the Claussen Stiftung/Stiftverband für die Deutsche Wissenschaft, whose support since 1987 has made possible a major part of her contribution to this Report.

Some Abbreviations

AID: Artificial insemination using semen from a donor.
AIH: Artificial insemination using semen from the husband.
CECOS: Centre d'Étude et du Conservation du Sperme. (Federation of sperm banks in France.)
GIFT: Gamete intra fallopian transfer.
IVF: *in vitro* fertilization (i.e. in the laboratory, rather than the womb.)
VLA: The Voluntary Licensing Authority, set up in the United Kingdom by the Medical Research Council and the Royal College of Obstetricians and Gynaecologists.
ZIFT: Zygote intra fallopian transfer.

Part One
Introduction

Introduction

When people look back at the twentieth century, the World Wars, or the rivalry between East and West in the second half of the century, may seem less important than they do now. Our century may be seen first as the time when we invented weapons that could wipe us all out, and yet managed to contain our quarrels long enough to survive. (On the other outcome, we will not be remembered). Perhaps our time will also be seen as the era when we became able to take control of our own biology, and in particular to take control of our own reproductive processes.

Ours is the first century in which many people have been able to plan one of the main aspects of their lives: the size of their families. The spread of contraception gave those who were fertile more control over their lives. But infertile people who hoped to have children were still unable to do so. Now we have developed, and are developing, techniques which extend our reproductive options. Among these techniques of fertilization two are central: artificial insemination and *in vitro* fertilization.

Artificial insemination is the transfer of semen into the woman's body by methods other than standard sexual reproduction. The semen may come from the husband (*AIH*) or from another donor (*AID*). AID opens up new reproductive possibilities, in that someone not intended to play any part in the upbringing is contributing to the creation of the child.

In vitro fertilization (IVF) is fertilization of the egg outside the woman's body. From this, various techniques have been developed. They include *GIFT* (Gamete Intra Fallopian Transfer) where eggs are mixed with the sperm and both are returned to the woman's fallopian tubes for fertilization to take place; and *ZIFT*, where what is transferred is a zygote. This family of techniques gives rise to such reproductive acts as egg donation, embryo donation, and the use of a woman's womb to produce a child for others (*surrogacy*). Associated with the development of some of these techniques has been *research* on embryos produced by them.

Medical science is giving us ways of reducing the incidence of congenital disorders. For instance *antenatal diagnosis* allows the possibility of *abortion* where the child would be handicapped. This already raises ethical dilemmas; and we are developing other ways of influencing the kind of children we have. Further down the line is the possibility of *gene therapy*: we might eliminate handicap without having to resort to abortion, by replacing the defective gene. Beyond this, we can envisage more dramatic developments: if we can replace defective genes, can we not also change other ones? This can be described either as a utopia in which we take control of our own evolution, or as a nightmare.

1. The aims of this report

This new set of problems, partly medical, partly ethical, and partly social, needs thought by many different individuals and groups. Any person who cares about the future of childbearing and of the family is inescapably concerned about these problems. Some infertile people, and some potential donors, have to decide what would be right for them to do. Scientists, doctors and members of related professions, either as individuals or through their professional bodies, find themselves required to work out views and policies. But the questions also require a response by society as a whole. What framework of law or other regulation should exist? How far should these decisions, affecting the future of the family and affecting which people are born, be left to emerge from the decisions of individuals or couples, and how far should there be a deliberately planned public policy?

These questions have already given rise to discussion at the national level. In most European countries there has been public debate, and some countries have set up commissions to propose public policies. Among the results have been the Report of the Benda Commission in West Germany, the Report of the Working Party set up by the Conseil d'État in France, the Report of the Santosuosso Commission in Italy, and the Report of the Warnock Committee in the United Kingdom. Some countries have introduced legislation.

In the European Community, the question arises how far the member countries should have a shared policy. As the barriers between the countries fade, it may seem increasingly absurd if, say, surrogate motherhood is banned in one place but allowed half an hour's flight away. And, if it is desirable to regulate embryo research, there is the problem of the inadequacy of trying to control scientific research by regulations which stop at national boundaries.

But, on the other hand, there are very different religious and ethical traditions in the different countries. People are unlikely to change their outlook on issues so fundamental because of some decision about harmonization made in Brussels or Strasbourg. An insensitive attempt to tidy up differences might, if it succeeded at all, obtain uniformity at a great cost in resentment. So the question arises: how far, if at all, should there be a set of Community goals and policies?

The European Commission asked this Working Party to think about these issues. The Working Party is made up of members of some of the relevant parts of the medical profession, and of representatives of other disciplines such as sociology, theology and philosophy (*see* p. viii). The members are drawn from half the countries of the European Community.

It would have been possible to bring in more members to represent more member states and more disciplines. But we decided that if we remained a small group we might stand more chance of making progress in our discussions.

The Commission asked us to consider ethical and other issues raised by the new reproductive technologies. This open-ended remit left us a lot of scope to interpret our task in the way that seemed to us most fruitful. The various national bodies with a similar task had a strong chance of their conclusions being translated into legislation. But in the case of a European report, this seemed less likely. Rightly or wrongly, laws on these topics are at present primarily the concern of national parliaments. We rejected the option of duplicating at the European level the work of the different national bodies. Alternatively, we could have investigated discrepancies between the policies of member countries, with a view to the Commission making proposals for harmonization in at least the less controversial matters. But we did not want to produce something boring and bureaucratic: an unread document about the small print of legislation.

All these questions about social policy depend on attitudes taken to the fundamental questions of value these technologies raise. We decided that the most useful contribution we could make is to attempt to uncover these fundamental questions. We have tried to see how far the evidence now available, both scientific and social, combined with rational argument about values, provides a basis for a defensible set of views and policies on these issues.

This committee has not always agreed on what it considered the right policies. We have tried to make a coherent set of recommendations wherever possible. But, where disagreements seemed deep, we have preferred to expose them, together with the reasons given on both sides, rather than to patch together some compromise. We think it contributes more to an understanding of the issues sometimes to present reasoned alternatives than to present an agreed position for which the real reason would be the need for unity rather than genuinely compelling arguments.

We do not expect to have provided definitive, or even generally acceptable, solutions to these problems. Others will often disagree with our views. We hope that this will at least stimulate them to work out more fully and explicitly what their reasons are. Different approaches can only be compared when they are worked out in detail. And only in this way can progress be made. We hope to have provided an analysis which will enable people to think more deeply and clearly about these matters, with a fuller understanding both of alternative points of view and of some of the implications of their own.

2. The questions

The introduction of new technology creates new possibilities and problems both for those directly involved and for society at large. The focus of this report is not on the techniques, but on their human impact. We look first at three main groups involved: parents, donors and children. In some cases they have different interests, and we develop an approach to the conflicts that arise. We also try to bring out the way these apparently 'medical' matters raise issues that go to the heart of the kind of society we want.

The next part of the report is about surrogate motherhood. We discuss conflicts of interest between the parents-to-be, the surrogate mother and the child-to-be. We consider the issues raised by proposals for commercial agencies in this field.

Next we discuss the issues raised by embryo research, together with other issues such as the possibility of using foetal organs for transplants. On the one hand, these fields are exceptionally rich in potential medical benefit. On the other hand, issues familiar from the abortion debate about the status and rights of the embryo or foetus are raised in this new context more sharply than ever before. We also have to think about the permissible limits of human experimentation. We try to analyse these problems, and suggest a possible approach to them.

Finally, we discuss the way these new techniques can enable us to influence the kinds of people who are born. Choices about the screening of donors of sperm, eggs or embryos affect who is born. Antenatal tests for handicap create the option of selective abortion. And so, perhaps raising even larger social issues, do tests for the sex of an embryo. Then there are issues raised by the possible future development of 'gene therapy' and other forms of 'genetic engineering'. We also try to contribute a little to the thinking about some of those issues, whose resolution may help determine the future shape of human life.

3. Some values at stake

The widespread concern about these new reproductive techniques is no doubt partly due to a resistance many of us intuitively feel to 'interfering with nature': to the substitution of artificial processes for natural ones.

Any general principle which rules out the new technology on these grounds does not stand up to scrutiny. Virtually the whole of medicine would also be ruled out. But there is a special unease related to this kind of intervention in natural processes. One expression of this is in the Catholic Church's document on these issues: the *Instruction on Respect for Human Life in its Origin and on the Dignity of Procreation*, issued by the Congregation for the Doctrine of the Faith in 1987. That document specifically makes the point that artificiality is not enough to rule out these techniques; but it also mentions 'the temptation to go beyond a reasonable dominion over nature'.

In part, the idea of a reasonable dominion over nature is to be understood in terms of religious views which are outside the scope of our discussion. But we believe that the Catholic document is also articulating concern which is shared by many who are not Catholics, and which is independent of any particular religious belief.

This unease perhaps stems from a number of sources. Having children is something uniquely important in our lives: it involves the privacy and intimacy of sexual love, and, at least in the ideal case, involves the couple together in the deepest and longest commitment they will undertake. We are uneasy when into this comes the coldness of a visit to a clinic or of a commercial relationship with a third party. And we are similarly likely to be worried by the separation of biological from social parenthood. What will all this do to how we see sexual relations and to how we see the family?

Questions about the future of the family are bound up with views about the position of women, and also with attitudes towards the possibility of homosexual couples bringing up children.

Surrogacy raises the problem of how far we can ask someone to go on committing herself to a biological reproductive role without reaping the benefits of parenthood. And the issue of paying donors of sperm or of eggs raises questions about whether we want a society which allows or restricts the extension of the free market and commercialization into these areas of life.

The issue of surrogate motherhood raises questions about how far there should be legal intervention of a 'paternalist' kind to prevent people committing themselves to an enforceable contract which may exploit them. On the one hand, there is the libertarian view that the law has no business interfering to prevent free citizens making whatever contracts they choose. On the other hand, there is the kind of objection that makes most societies rule out contracts where people sell themselves into slavery. Large issues are raised about the limits of state intervention to protect people from the harmful effects of their own decisions.

The question of embryo research is partly a matter of what rights or interests are possessed by the embryo, and the possible conflict with the interests of these who may benefit from research. But it also raises issues which transcend those conflicts: questions about the wider social impact of such experimentation and about the social control of science.

It is not surprising that people have conflicting responses to many of these issues. They touch on our attitudes towards sex, marriage, parenthood, and handicap. They are also linked to more general features of our outlook, where we may be influenced by different beliefs about religion, about the position of women, or about children's rights. They raise in addition some of the issues in the abortion debate, questions about the social control of science and technology, and questions about the limits of state intervention to regulate what people do with their own lives. With all *these* issues at stake, it would be astonishing if there were not passionate disagreements about what should be done.

The deep divisions over present developments are partly about whether they are in themselves acceptable. But partly they reflect concern that we may be sleepwalking, step by step, into a world which few of us would now choose. These techniques are helping to shape the society we live in, and in turn their development is influenced by the social choices we make. Because of some of the possible longer-term implications of these developing technologies, a coherent social response to them is exceptionally important.

Part Two
Parents, Donors and Children

Chapter 1
An Approach to the Problems

In thinking about IVF and similar techniques, parents, donors and children are the main people to consider. Many of the problems concern who they should be; that is to say, the questions are about which people will become parents through these means, and about the recruitment of donors. And the answers to these questions will influence which children are born.

Other problems concern the kinds of relationship there should be between parents, donors and children. Some of the most difficult questions arise when there are conflicts of interest between any of them. And some of these questions affect society as a whole, by affecting the way we see parenthood, children and the family.

1. Problems: donor recruitment and anonymity

The main questions here are:

Should the identities of donors (of semen, eggs or embryos) be kept anonymous?

If donors should *not* be kept anonymous, is it better if they are relations or friends of the prospective parents, or people previously unknown to them?

Should donors be screened?

Should donors be paid?

2. Problems: access to help in having children

The 'standard' case is that of an infertile married couple seeking help in having children. But there are other cases. An unmarried couple, or a single person, or a homosexual couple, may also seek to have children by these means. Should help be readily available to all these prospective parents, or should access be restricted? And should there be screening on other grounds? For instance, if one of a married couple is seriously mentally handicapped, or is an alcoholic or a drug addict, should this exclude help in overcoming infertility? Or what about a woman who is fertile, but who wants the help of a surrogate mother because, for instance, pregnancy would interfere with her job?

3. Conflicts

Some of these questions are difficult because there are conflicts of interest.

In the case of semen donor anonymity, the child may want to know his or her biological father. But the man may want to forget all about having donated some sperm years ago, and the social parents may not want their family complicated by a relationship with the biological father.

Or there may be a conflict between the interest of the single person or the alcoholic who wants a child, and the interest of a child in having two parents without severe problems. (This conflict is complicated by the intrinsic difficulty of deciding who will be good parents. Moreover, the choice is usually not whether a particular child will have one set of parents or another, but whether that child will be born at all.)

We look first at ways of thinking about these conflicts. Our discussion here of these approaches applies not only to parents, donors and children. The same underlying issues arise in other conflicts, for instance between the possible interests of the embryo and the interests of those who may benefit from embryo research, and at other points throughout this report.

4. The utilitarian approach to conflicts of interest

One approach is the utilitarian one. Everyone's interests are taken into account, and weighed against each other. The right policy is determined by how many people stand to gain or lose by one decision or the other, and by how much the gains or losses matter to them.

Take the case of whether sperm donors should have their identity revealed to their biological children. Utilitarian thinking about this starts by trying to assess how serious a loss it is not to know your biological origins. This loss to the child is compared with the disadvantages to the donor of later contact with offspring he might rather forget about. The ideal here would be to be able to consult substantial numbers of people who had two characteristics: being products of unknown sperm donors, and also being donors who were known to (and had been approached by) those their donation had produced.

The donors and their offspring are not the only people for a utilitarian to take into account. If lack of anonymity reduces the numbers of donors, some potential parents will be denied children. Their deprivation has to be included in the assessment. And, more controversially, some versions of utilitarianism would include the interests of those who will be born only if there are more donors.

It is obvious that in real life much of the evidence ideally required is not available. So the utilitarian 'calculation' has to depend on a mixture of indirect evidence and intuitive assessment. The utilitarian may regret this limitation, but may not see it as a devastating objection to the approach. Moral decisions simply may not be susceptible to precise scientific methods; and any plausible moral view gives some weight to the effects of actions and policies on people's interests, and so shares with utilitarianism this limitation.

An objection more specifically directed at utilitarianism is that it disregards people's rights. It is often said that, for a utilitarian, 'the end justifies the means', and that pursuit of the general good may mean riding roughshod over individuals. On the issue of donor anonymity, it may be held that the child has a *right* to know his or her biological origins. If there is such a right, the need for donors does not justify denying the child this knowledge. If the utilitarian supports anonymity as the policy with the best consequences, critics will see this as a case where utilitarianism gives what is morally the wrong answer, and does so by treating a right as merely another interest to be considered.

Utilitarians have in general been sensitive to criticisms of this sort. Refined versions of utilitarianism have been produced in response to them. One such version is 'rule utilitarianism', according to which the consequences of individual acts are not what matters. On this view, we should calculate the consequences of obeying different rules, choose the best set of rules and use them to judge individual actions. Another is the 'two level' theory of R.M. Hare, where at the 'intuitive' level we should act on rules or cultivated dispositions, while at the 'critical' level we should evaluate the rules or dispositions on the basis of their utility (R.M. Hare: *Moral Thinking, Its Levels, Method and Point*, Oxford, 1981).

Consider for a moment the utilitarian approach to some issues outside the field of reproductive ethics. Certain interests have a special importance for us: these include life, freedom from arbitrary arrest, freedom of speech, and not — except in clearly defined circumstances and with our consent — having our lives put at risk by doctors carrying out medical experiments. It has been held that people will only feel secure if there is some guarantee that these vital interests will be protected. Many utilitarians have, on these grounds, been prepared to justify the adoption of rules protecting these interests, such as the principle that research should never be carried out on patients without their informed consent, even in circumstances where other interests appear to outweigh the need for security.

A sophisticated utilitarian can recognize that limitations on the scope of utilitarian calculation can themselves be justified on utilitarian grounds. Although utilitarians do not believe in absolute rights, this approach produces a *de facto* recognition of frontiers which closely correspond to rights.

These oblique strategies are themselves controversial. One question is whether they generate a stable position which is still utilitarian. Perhaps the rule against risking patients' lives in medical experiments which will not benefit them is based on people's need for security. This consideration may often outweigh the short-term utilitarian case in support of such experimentation. But what is to be done in the extreme case where the utilitarian calculation may go the other way, even if the long-term considerations are included? (Perhaps the experiment would save huge numbers of future lives.) In such a case, to stick to the rule is to abandon utilitarianism. And to stick to utilitarianism is to violate what others take to be a right. Those who believe in such rights may feel that utilitarianism does not provide a secure enough basis for them.

5. Rights, dignity, respect for persons

An alternative approach centres around the idea that people have a certain dignity, or should be treated with a certain respect. An ethic based on respect for persons sometimes finds expression in Kant's phrase that people should be treated 'always as ends in themselves and never merely as means'. This gives a reason for objecting to the medical experiment, even if calculation of consequences comes out in favour of it. On this view, the doctor may not simply look through the patient to the future beneficiaries of the research. The patient has a claim to more respect than merely having his or her interests given equal weight with those of any one of the potential beneficiaries.

It would be too stringent to require that no-one's interests are ever put second to those of others. That would rule out any solution to most cases of conflict of interests. The more plausible interpretation is that certain interests are too central to be sacrificed: in other words, that they generate certain rights which should not be violated.

One problem is that this seems to give us the framework of an approach to moral problems, without its substance. Everything depends on what rights we have; and more fundamentally, on what procedure we use to determine what rights there are.

Rights theorists disagree both over what rights there are and about the justification for claims about rights.

There is also a problem about whether rights should be thought of as absolutely inviolable. For instance, there are conflicts between different rights. Suppose we think that a child has a right to be born into a family of a certain kind, perhaps a family with two parents. A single woman gets pregnant. Are we to say that she has a duty to have an abortion, whatever her own wishes or convictions? Or does she too have rights to be considered? The content of the child's right is indeterminate until such questions are answered. If the woman has relevant rights too, some order of priority seems necessary.

Another aspect of the problem of whether rights are absolutely inviolable has to do with the priority that rights take over other considerations. Suppose we say that the child resulting from AID has a right to know his or her biological parents. Would this right have to be respected even if so many donors were deterred that the infertility programme collapsed?

In an attractive image drawn from card-playing, Ronald Dworkin has suggested that rights are trumps (*Taking Rights Seriously*, 2nd impression, London, 1977, page xi). This captures something distinctive about rights, but brings out a certain rigidity in rights theory. It may seem procrustean to suppose that all moral claims we care about can be sorted into the two categories of rights and non-rights, such that the least right trumps any claim in the other category, no matter how substantial. We may become cautious about recognizing rights until we have thought out the extent to which we may be boxed in by them, unable to avoid a disaster because the only way of doing so involves some rights violation of minor importance.

Rights theorists have in general been aware of these problems. They have developed sophisticated theories, with various rules of precedence between rights. And they have produced accounts of *prima facie* rights, which can be overridden in circumstances of great necessity. A difficulty for sophisticated forms of rights theory is whether they can keep to the fine line between the perhaps excessive flexibility of utilitarian theories and the perhaps excessive rigidity of the simpler rights theories.

6. Convergence

We have seen that neither the utilitarian approach nor the rights-based approach is free of problems. It is not yet clear that there is a stable utilitarian position which gives individuals the kind of protection which would generally be accepted as adequate. And it is not yet clear that believers in rights can provide a satisfactory justification for drawing the distinction between rights and other interests in one place rather than another (or that they have yet provided a satisfactory account of how conflicts of rights should be adjudicated). If utilitarianism seems alarmingly flexible, the two-tier approach of rights theory seems disturbingly rigid. This central region of ethical theory is one of problems still unsolved.

There are several possible explanations for why this might be so. Perhaps both approaches are fundamentally flawed, and some as yet undiscovered third approach is needed. Or, it may be that people have irreducibly different values, so that the search for the theory which will seem satisfactory to us all is misguided. Or, it may be that some further improved version either of utilitarianism or of a rights theory would escape the problems of current versions.

Simple versions of utilitarianism and of rights theory are clearly opposed. They each have difficulties the other avoids. Sophisticated versions of each try to incorporate some of the strengths of the other. But in doing this, they tend towards convergence. The problem for each is to incorporate the strengths of the other while retaining its own identity. While this may be a problem for theorists, those of us trying to solve practical ethical problems may be encouraged. Perhaps a plausible set of solutions is to be found in the area of convergence between the sophisticated versions of the two approaches. This area seems worth exploring regardless of which label turns out to fit better the most acceptable approach.

We do not as a committee feel able to solve these central

issues of ethical theory. In the absence of such solutions at the theoretical level, perhaps the best that can be hoped for is to get as close as possible to what John Rawls has called 'reflective equilibrium' (John Rawls: *A Theory of Justice*, Harvard, 1971).

As Rawls describes this, it is approached by formulating general principles which seem plausible, and then seeing to what extent their application fits our intuitive responses to particular cases. Where there is conflict between intuition and theory, we need to reconsider both. We need to ask whether the theory should be modified to accommodate the intuitive response. And we also need to ask whether the intuition is one which, on reflection, we are prepared to abandon. The hope is that, by a process of mutual adjustment, we may reach a state of equilibrium, where we have a stable set of principles and of intuitions, which are in harmony with each other.

There are, of course, questions about how far an individual person's values are susceptible of being harmoniously systematized in this way. And there are perhaps even bigger problems about the emergence of any social consensus from this process. Even if each of us reaches our own reflective equilibrium, perhaps the resulting views will diverge radically from each other.

We take it that what degree of consensus will emerge is an open empirical question. Particularly in a new field of ethics of the kind we are dealing with, it is hard to be sure to what extent disagreements are mere surface phenomena, the product of insufficient experience and thought, or expressions of deep differences of values.

Our approach here has been to consider the interests of the different people involved with a certain flexibility. We do not have a prior commitment to the idea that simply summing utilities will give the best answers. But nor do we feel committed to the view that any of the parties has rights which may never be overridden. In this way our approach is somewhere in the middle ground shared by sophisticated versions of utilitarianism and of rights theory, from both of which we have borrowed something. Perhaps one way of mapping that middle ground is to look in detail at such particular conflicts as we are concerned with here. If intuitively acceptable resolutions of them can be found, these may in turn help us to evaluate claims made on behalf of more general theoretical views.

Chapter 2
Donor Recruitment and Anonymity

1. Recruitment of semen donors

For many years, it has been common to ask medical students to donate semen, usually for payment. This approach has been questioned. It can seem a socially based eugenic policy: physicians are choosing young members of their own profession as being the most intellectually and physically fit candidates for semen donation (G. Annas: 'Artificial Insemination: Beyond the Best Interests of the Donor', *The Hastings Center Report*, (1979) 4: 14–15, 43). There is also a tendency with this type of recruitment to use a few 'good' donors repeatedly, thus creating a greater risk of marriages between genetically linked AID children. And if the request comes from a more senior member of the profession, able to influence the student's career, this could create undue pressure to consent.

Semen donation is not just like blood donation. By donating semen for these new techniques, a man is partly responsible for bringing a new person into the world. The potential donor needs time to consider his motives, and possible future regrets. Perhaps a donation made by a young unmarried man is something he will later find difficult to talk about to his wife and children. At some centres, especially in France, potential donors first have a long interview. This should ideally give them a chance to think, and some protection against any strong pressures from family or friends to donate.

An alternative to using medical students has been developed by the French Federation of CECOS banks. They appeal for donors through the media. They also ask the recipients themselves to approach family and friends. (These donations will not be used for them but for other anonymous recipients.) Their ideal is a married man with healthy children, donating with his wife's consent; and there is no payment: semen donation is thought of as 'a gift from one couple to another'. Moreover, no donor's semen is used for more than five pregnancies.

CECOS policy has diversified the donor population. Donors have a wider age span and a wider range of social backgrounds. However, because it is based on an attitude of greater openness about infertility and AID, it can conflict with the desire of recipient couples for privacy, or with misgivings among potential donors about the morality of AID. CECOS banks have in fact found it hard to keep up the supply of donors to meet increasing demand.

Other countries, such as West Germany, discourage AID altogether, partly because of its eugenic aspects, and partly because of problems about anonymity.

There are different policies about payment: most often donors are paid, and the French CECOS banks have been told more than once that they could overcome their donor shortage in this way. A study done by one of the French semen banks at Necker hospital in Paris does support this view. The bank, which originally paid its donors – mostly single students – had to stop paying them in order to join the CECOS Federation. They were, however, allowed to continue recruiting single men, essentially for comparative research purposes. A comparison of the two donor populations in 1980 did show a decline in the total number of donors after payment was stopped. However, the mean age of donors increased, the number of married donors doubled, and the number of white collar donors tripled

(C. Da Lage, M.O. Alnot, P. Granet, G. de Parseval: 'Les Donneurs de Sperme', in Conseil Supérieur de l'Information sexuelle, de la Régulation de naissances et de l'Éducation familiale (eds), Colloque International: *Les Pères Aujourd'hui* Paris, INED, 1982.)

Payment therefore results in more donors, and perhaps some men do donate semen for money. But there may be something more subtle about this. Men may have other reasons for being willing to donate, but the payment gives them a 'reason' which damps down further self-scrutiny of their motives.

Donors may have conflicting interests about payment. It can be seen as social recognition of their action. But it may also remove some of the sense of having acted altruistically.

2. Screening semen donors

Some medical screening is not seriously controversial. Donors should be screened to exclude carriers of venereal diseases or the HIV (Aids) virus, and carriers of inborn disorders. Screening, accurately recorded, has obvious medical justification, and where AID is highly institutionalized, as in hospitals and sperm banks, it usually goes beyond physical examination and medical history, to include laboratory tests: Rhesus factor, blood group, semen culture, etc. To avoid combining recessive genes, recipients are sometimes screened too. With new techniques available, more tests are likely to be thought necessary for the best chance of a healthy child.

More controversial is the attempt to match donors with social parents in race and general appearance. Normally only medical and morphological criteria are adopted: blood group and Rhesus factor, and colour of the skin, hair and eyes.

A major future issue will be the boundary between acceptable 'medical' criteria and unacceptable 'eugenic' screening. There may be no serious problems about selecting donors broadly like the social parents, despite some unease about how far this should go. And there is likely to be little objection to screening for severe disorders. But, with more refined techniques, screening for more subtle abnormalities may look more like a slide towards eugenics. This boundary problem is one we should start thinking about now.

3. Should semen donors be anonymous?

Most semen donors express a preference for anonymity. This is partly to avoid paternity suits. (In most countries, the legal position is unclear.) But it is also to avoid unwanted later contact with their 'offspring'.

Swedish law has given the child the right to know the identity of the semen donor, on reaching maturity at the age of eighteen. Paternity suits are eliminated by assigning paternity to the married woman's husband, who gives his irrevocable written consent. The donor remains anonymous as far as the social parents are concerned. The law equates AID with adoption, so the donor has a socially recognized position, though one without rights. Is this alternative model preferable to anonymity?

Let us look first at the family in which the child will grow up. Social parenthood often out-ranks biological parenthood. Being a social father is much more important in life than being a semen donor. And the emotional bond with the social father is usually far more important to children than the genetic links with the donor.

The social parents may want their family to be a closed unit, as much like other families as possible, unencumbered by ambiguous half-relationships with donors. A social father may feel rejected if he sees the donor as a rival.

Parents often prefer anonymous donors who will disappear afterwards. But some opponents of anonymity favour the Swedish model, where the potential identification of the donor would only take place eighteen years after the child's birth. This seems to give plenty of time for the development of family bonds which will survive. And parents themselves sometimes prefer a known or related donor. This can be because they think they have some idea of the likely genetic characteristics of the child. And, in the case of related donors, they may value the extra genetic link with the child this gives, as well as their more intimate knowledge of the kind of person the donor is. It is possible to have a known or related donor whose identity is kept from the child; but this involves the drawbacks of family secrets. The desire for a genetic or other link may lead some social parents to prefer a system without anonymity.

What about the position of donors? As in the Swedish system, the donor can be given complete legal protection, so that the child has no rights against the donor other than knowledge of his identity.

The effect of abolishing anonymity in Sweden seems to have been an initial decline in numbers of donors. This may suggest that many donors prefer to be anonymous, quite apart from fear of paternity suits. But this must be linked to two other effects of the new law. There was a decline in demand: couples felt less comfortable at the thought that the child might eventually wish to contact the donor. And physicians in some AID centres refused to continue offering AID under the new law. In the centres still continuing with AID, the numbers of donors have returned to normal, although they are now more often older and more often married (G. Ewerlöf: 'Swedish Legislation on Artificial Insemination', typescript; M.A. d'Adler and Marcel Teulade: *Les Sorciers de la Vie*, Paris, 1986, Part 2, Chapter 4, pages 127–144).

Policies on anonymity represent a social choice about the meaning of donation. Do we accept and recognize the donor's contribution as an act of altruism, perhaps as part of a system in which a husband donates with the full agreement of his wife? Or do we prefer the anonymous student as a source, treating the contribution as an embarrassment, to be accepted but swept under the carpet? There may be more dignity for the donor in a system of openness rather than anonymity. In the case of donors, there seems something to be said for eliminating anonymity, but against this must be set what appears to be their own widespread preference for retaining it.

What about the children? The child's concern with his or her origin was the main motive for the Swedish policy.

Some adopted children find they come to care very much who

their biological parents are, and may go to great lengths to find out. Our sense of who we are is bound up with the story we tell about ourselves. A life where the biological parents are unknown is like a novel with the first chapter missing. Also there are the marked similarities between children and their biological parents. The child may wonder who is the person, perhaps among those passed in the street, who has that degree of closeness.

On the other hand, for young children who know who the semen donor was, there may be problems about their identity. They may see neither person as being unambiguously their father. This suggests that it may not be in the children's interest to be told who the donor is at an early age, but is not a point against a system of the Swedish type, setting the right to know at the age of eighteen. And, since the legal right to know need not be exercised, no child loses anything by it. Since some people care so much about their origins, seeing them as an important part of their identity, the interests of the children count strongly in favour of the right to know.

What relative weight should be given to the different interests of the various people involved? This is the kind of problem where no absolute general rule is likely to give best results in all cases. So much depends on the individual case: who the particular people are, their relationships and what they care about. As a committee we unsurprisingly found ourselves differing in the weight we gave to the different interests.

Some of us felt that it can be very hard on the parents to have the donor intrude on the family. But most of us were inclined to think that, by the time the child is eighteen, the family should usually be strong enough to weather this.

Some of us were inclined to see knowledge of one's origin as so central to identity as to be a right. We all accept that ignorance of it can be a severe psychological disadvantage, and we give this great weight in thinking about policy. But the claim that this knowledge is an absolute right suggests, for instance, that it should always outweigh any degree of unwillingness by donors to discard the protection of anonymity. Is this plausible?

In a system without anonymity, donors need not themselves be hugely disadvantaged. As in Sweden, their legal position can be protected. There may be some disadvantages in later contact by their offspring. But no-one need become a donor if they think this possibility is a terrible one. Perhaps the interests of the children count for more than the possible disadvantages to the donors.

But the case for anonymity does not here simply rest on a direct appeal to the interests of the donors. The fear is that, through putting off potential donors, abolition of anonymity will damage the whole programme. The losers will be infertile couples who will no longer be able to have this help in having children because potential donors have voted with their feet.

The extreme views are, on the one hand, that knowledge of the donor is an inviolable right, and, on the other hand, that anonymity should always be guaranteed. Perhaps a reasonable middle course can be found.

We suggest that the child's interests create a strong *presumption* in favour of openness, but with protection for the various parties involved. As in the Swedish model, the social parents should be protected from intrusion when the 'child' still *is* a child, and the donor should be protected from paternity claims. But, although we favour openness, this is a presumption rather than an absolute right. There is a case for adopting a Swedish-type law for an experimental period, and seeing what happens to donor recruitment. If it slumps disastrously, public appeals could be tried to counteract the effects of the new system. If none of this worked, there would then be a case for abandoning the experiment.

To put the point briefly: it can be better for a child to be born without the right to know the biological father than for that child not to be born at all. But, if the donor programmes can be kept up, best of all might be to be born with the right to know.

4. Donors of eggs and embryos

The legal status of the egg donor is unclear. Egg donation raises the same issues of social attitudes as sperm donation. However, many people do see the two kinds of donation as different. (For instance, the West German Medical Association bans egg donation altogether on the grounds of the interests of the child-to-be. Yet AID, while generally discouraged, is allowed in exceptional circumstances.) One thing both forms of donation have in common is that our societies have not decided whether they should be concealed or brought into the open, and the undefined legal position reflects this.

Unlike semen donation, egg donation is often risky. Laparoscopy, until recently the normal method of egg retrieval, involved surgery under general anaesthesia. Newer ultrasound techniques approach the ovary via the vagina, the bladder, or the peritoneum. They do not need surgery, but carry some risk, and can be rather painful.

In Europe, there are no public appeals for egg donors, and they are usually unpaid. It is generally supposed that only a woman close to the recipient is likely to accept the trouble and risk. Couples sometimes bring a potential donor to the first consultation. Because of the emotional links, there is a question of how to protect unwilling women from being pressured by family or friends into donating eggs.

Where a couple are both infertile, donation of a pre-embryo rather than an egg is necessary if the woman is to bear a child. This is like adoption, but with the experience of pregnancy and birth, together with any bonding which may take place as a result.

There are two differing problems. One is that of finding the method which carries the least risk of harm to any of those involved. The other is that of deciding who will donate the embryo. This second question raises a cluster of questions about the relationship. For instance, who, if anyone, 'owns' the resulting embryo? Should the donating woman — or the donating couple — have a veto on what happens to the embryo? Or can their wishes be overruled to help an infertile couple where there is a shortage of eggs?

5. Eggs and embryos: known and unknown donors

So much more is involved in donating an egg than in donating semen that it is usually undertaken by someone who is close to the woman who wants a child. Friends or members of the family may be willing to donate as an act of generosity. Is it better that the donor should be a relation than that she should be someone unknown?

The childless couple often prefer a related donor. They value the fact that this gives one of them a genetic link with the child. They may prefer to have *some* idea of the genetic characteristics their child may have. And they may like the atmosphere of an act of generosity within the family, thinking this better than an impersonal or even commercial transaction. (There is a parallel with the case of surrogacy, where in West Germany the report of the Benda Commission considered a legal ban on anyone *not* a relation being a surrogate mother.)

The lack of anonymity in egg donation could generate future rivalry between the two women. The CECOS policy for semen donors gives one alternative model. In some French hospitals, donors are exchanged so that the recipient is given an egg from an unknown donor. But this policy will not suit potential parents who prefer the egg to be provided by, for instance, a sister.

The case for unknown donors is based on the interests of the child. Where the donor is in the family, this fact can either be revealed to the child or not. It is suggested that either policy can be bad for the child.

If the child is not told, this means that the central relationships of childhood are founded on evasiveness and deception. For a family to have this kind of secret at its very heart may be corrosive of relationships. Also, this kind of knowledge is so central to a person: perhaps others have no right to hold it back. And it may come later; not everyone can keep a secret for ever. There may be a time when it has to come out. For instance, when the child grows up, a family history may be necessary to determine genetic risks to his or her own children. To discover later in life that you have been lied to about where you belong in your own family can be devastating. The case against this kind of secrecy is very strong.

The alternative is to be open with the child about his or her genetic mother. But this may lead to problems too. Perhaps being told that your social aunt is your genetic mother gives a blurred picture of family relationships, leading to a confused sense of your own identity.

The strength of the arguments against secrecy and deception is enough to rule them out. The choice should be between stipulating unknown donors and being open about having a related donor.

The Voluntary Licensing Authority set up in the United Kingdom has said that 'egg donors should remain anonymous and for this reason donation for clinical purposes from any close relative should be avoided'. It is not clear from this statement, which is not further elaborated, exactly what reason the Authority has in mind. It could be that they would regard a related donor as acceptable if secrecy could be guaranteed, but suppose (plausibly) that no such certainty obtains. If so, their view of what is desirable in the case of related donors is the opposite of ours. Or, possibly, they agree with us that secrecy in the family should be excluded, but also think that anonymity should here be an absolute priority. We agree that the combination of these claims logically excludes related donors, but we wonder whether the Authority has sufficient reason for placing this absolute value on anonymity.

When secrecy is excluded, the central issue is this: how far do possible identity problems outweigh any parental preference for a related donor?

There is little clear evidence about the extent of these identity problems. The claim seems to be largely conjecture. This is unavoidable with techniques so recently put into use. One need does seem very clear: there should be a follow-up study of very large numbers of the children born as a result of these new reproductive technologies. We could then see the effects of these different approaches. In the next generation, decisions about them would not have to rely on such a large ratio of speculation to evidence.

The conjecture about related donors causing identity problems may be true. And, the possible parallel with adoption suggests that there may also be identity problems caused by *not* knowing these important facts about your origins. At present there is no basis for assessing the relative severity of the problems in the two kinds of case. So there is not enough reason for any general exclusion of donors of either type.

Having said this, we have some reservations to express. The standard and most discussed case is that of donation by a sister. And cousins, for instance, sometimes donate. But some other donations may cause disquiet. In one case in the United Kingdom, it was proposed to transfer an egg from an eighteen-year-old girl to her mother. This was stopped by the hospital's ethical committee. (But a similar case, involving surrogacy rather than egg transfer, has taken place in South Africa.) Fertilization of such an egg by the girl's father would bring into play various factors, including the increased risk of genetic disease, which lead to the prohibition of incest. Apart from this, the proposed egg transfer might be thought disturbing for two different reasons. One is that there seems a peculiarly deep blurring of the normal family relationships in such a case. And the other is the problem of being sure that a daughter's consent is fully voluntary. (In the case in question, the daughter was about to be given away in an arranged marriage.)

It is hard to evaluate the unease about the deeper blurring of relationships in a daughter-to-mother transfer. It may be that such feelings are just an irrational resistance to things that disturb our traditional categories. Or it may be that these feelings are symptomatic of some deep feature of our nature, such that tampering with these relationships will do great psychological harm. Those who take the first view will see daughter-donors as merely an extension of what is already happening with sister-donors. Those who take the second view have a reason for thinking some of the more startling cases of related donors should be excluded.

The other question, of how genuinely voluntary the consent

is likely to be, is disturbing in a less debatable way. It may be hard for daughters to say 'no' to the desperate desires of mothers. Perhaps there should even here be no absolute exclusion: there could be cases where the consent was genuinely free, and where the absence of other donors would mean that such an exclusion would prevent the birth of a much wanted child. But the difficulty of being sure that consent is fully free suggests that there should be a strong presumption in favour of donors other than daughters when they are available.

There is a further issue worth mentioning. Should it be up to the parents to choose between known and unknown donors? Or should there be some decision on this taken by the hospital or by society as a whole? This is a case where the presumption in favour of parental choice could possibly be overridden if there appears good evidence that one or other is substantially better for the children.

6. The availability of unknown donors

While some potential parents prefer egg donors to be relations or friends, others think unknown donors raise fewer complications. But such a preference can only be met if anonymous donors are available.

One source for donation can be a woman having IVF, in a case where the ovulation of an unusually large number of eggs is deliberately brought about. Some of the eggs resulting from this superovulation are given to another woman. This has the added advantage of making surgery unnecessary. But it may be hard on the woman who donates the egg, if the donated egg leads to a baby and if her own treatment is unsuccessful. Another possible source is women undergoing voluntary sterilization. As long as the procedure is safe, and consent is not given under pressure, this is a promising way of finding donors.

Chapter 3
Parents and Children

Sometimes the problems raised by these ways of helping people have children seem so great that it is tempting to think we would have been better off if they had never been developed. This is a possible view, but it is one which should not be seriously adopted before looking at two groups who can be said to benefit from them. These are the parents and the children.

Those who favour ready access to these techniques sometimes invoke a 'right to procreation'. But where setting up a family has been declared a human right (as in the United Nations' *Universal Declaration of Human Rights* and in the *European Convention on Human Rights*), the intention appears to have been to protect natural procreation from interference. A right to help with reproduction cannot simply be inferred from the right to set up a family. We are also sceptical about deriving views about what to do from abstract declarations, and prefer to start from the interests of the people concerned.

We start with the parents. The intuitive idea of a parent fuses together a biological and a social relationship. Historically, the two have gone together, with a few exceptions such as adoption. But the new techniques, separating the biological from the social, give the word 'parent' further ambiguity. Here we are considering those who are, or who hope to be, the social parents of the child. The first thing to note is the variety of reasons, not all of a medical kind, which can lead them to seek help with having children.

1. Seeking reproductive help

(a) Infertility: The simplest motive arises where a couple are unable to have children. Infertility, affecting about one in ten couples, is easy from outside to under-rate. But its psychological effects can be devastating. The couple may feel shame and loss of self-esteem. It may lead each of them to self-recrimination or to recrimination against the partner. Through the psychological association of procreative potency with sexual potency, it can cause sexual problems. It can also cause marital disharmony, divorce, and sometimes social withdrawal and even suicide.

Few would oppose medical treatment for infertility. But is using IVF a form of medical treatment? It is not a treatment in the way that giving hormones is. It does not restore the woman's capacity to conceive. It is a method of fertilization (requiring medical skills) which replaces 'inefficient' sexual relations. Because the medical profession makes available this alternative to sexual relations, reproduction now sometimes involves a 'professional' third party, and sometimes donors, as well as the couple who want a child. One of the deepest issues raised is how far having children should move away from being an intimate family matter; and there is a danger of this issue being obscured by thinking of the issue in terms of medical treatment.

The question of whether the 'medical' model is appropriate is important to clear thinking about the issues. But whether or not IVF counts as a form of medical treatment is not decisive for its acceptability. What matters is how it affects those involved, together with its implications for society as a whole. We do not wish to make the morality or availability of IVF depend on a definitional issue.

Sometimes adoption is an available alternative. But often what people want is not just to have a child, but to have *their* child: one biologically linked to them, either by genetic inheritance or by the 'mother' having given birth to the child.

Another reason for wanting to use the new reproductive techniques stems not from infertility itself, but from the 'relative infertility' of there being special medical risks for the mother-to-be attached to conception. (A case of using the new technology to circumvent a different sort of risk is where parents whose children will have a high risk of handicap, and who might have been reluctant to conceive, go ahead because they can opt for abortion if antenatal tests reveal abnormality.)

(b) Non-medical Reasons: While the central case of demand for these techniques may be that of the infertile heterosexual married couple, the range of possible 'parents' is much wider than this. A professional woman might want a surrogate mother to avoid her career being interrupted by pregnancy. A single woman might seek AID. Or a widow might want to be impregnated with her husband's stored semen. Some lesbian couples have been helped to have children by AID.

The boundary between medical and non-medical reasons may again be blurred. A woman who, after several miscarriages, looks for a surrogate mother, is perhaps a borderline case.

2. Possible drawbacks for the parents

There can be drawbacks as well as benefits. There may be a conflict with other values. At one extreme, techniques using donors have been described as a form of adultery. (The Catholic Church prohibits IVF, both because it is morally wrong in itself and because of its effects on values. The Catholic Church does, however, accept the use of the GIFT technique, subject to conditions: among them that the semen be collected by means of a permeable condom during intercourse, so that neither masturbation nor contraception is involved.) Even without moral conflict, there may be emotional resistance. In AIH, the husband may be troubled by having to obtain the semen by masturbation. The woman may have a psychological reaction after impregnation. Infertile husbands may feel guilty.

Perhaps a husband and wife will start to feel alienated from each other. It has been suggested that subsequent adultery might be encouraged.

Some of these worries may be exaggerated. In France couples who have had children by AID have fewer than average divorces. (Christine Mannel: 'Le père de l'enfant né par IAD', *Troisième Congrès de la Société d'Andrologie de Langue Française* (in press)). The long waiting list for donors may deter most couples who are doubtful whether they can handle AID. (The two most common ways in which things do go wrong are the husband rejecting the child, and the marriage breaking up.)

Some have seen disadvantages for women in these techniques. Perhaps reluctant infertile women will be pressured into using them. Feminists have suggested that an infertile woman's desire for children is itself partly shaped by the social pressures towards motherhood, and that reproductive technology enables the (predominantly male) medical profession to take control of this central part of women's lives, in part by fragmenting motherhood. One feminist writer has asked: 'Why are they splitting the functions of motherhood into small parts? Does that reduce the power of the mother and her claim to the child? ("I only gave the egg. I am not the real mother." "I only loaned my uterus. I am not the real mother." "I only raised the child. I am not the real mother.")' (G. Corea: 'Egg Snatchers', in R. Arditti *et al: Test-Tube Women*, London, 1984).

Fatherhood is also fragmented by reproductive technology. (This seems less evident in English than in French, where the term 'géniteur' (biological father) can be distinguished from 'père' (social father). In English, the expression 'to father a child' links paternity to the act of conceiving the child. In French law, despite some recent changes in favour of the biological father, the main presumption is that the father is the mother's husband. English law gives more emphasis to the concept of illegitimacy, perhaps reflecting the linguistic tie between conceiving and fatherhood.)

In some cases a conflict between the parental couple may develop. Tests showing the presence of handicap may divide them over the question of abortion. Or they may disagree about whether or not an embryo should be implanted.

3. The interests of the children

The children are those most deeply affected. Their family circumstances may be unusually complicated: like adopted children, they may have to get used to the fact that at least one of their social parents is not their biological parent. Much more importantly, their very existence results from reproductive help. If they are glad to be alive, they are perhaps the greatest beneficiaries. They are the people who have no say at the time the decisions are taken, and their needs and interests have to be given great weight. Their central need is for a certain kind of home and family.

Some things are uncontroversially better for the children than others. With surrogate motherhood, it is obviously bad for the child if there is a long battle for possession. It is also obviously important for a child to have social parents who will bring him or her up. Very few people indeed would support using IVF and then letting the foetus develop entirely in the laboratory, without any parents waiting.

Things are less clear when we ask what kind of family and home is best for the child. It is hard to know who will be a good or bad parent. But children need warmth and love. Some of the worst cases of cruelty to children would make any humane person wish that those parents stayed childless. It is surely unthinkable that anyone would have wished to help them have children. Of course we cannot easily predict such disasters. But, where there seems a risk, there is a strong reason not to help.

Some other cases are much more controversial. Is it a disadvantage for a child to grow up with only one parent? (This often happens through death or separation, but should it be seen as a situation we ought deliberately to bring about for a child?) Is it a disadvantage to be brought up by a lesbian couple rather than by a mother and father? What about parents with severe psychological problems, say where one parent is an alcoholic or a drug addict? What about parents who are severely mentally handicapped? In some cases, girls with severe mental handicap have been sterilized, because they were thought unable to bring up children, and some have said that this infringes civil liberties and is discrimination against the handicapped. But it could be said that refusing to help someone have a child is different. A lot depends on the weight given to the difference between preventing someone doing something and refusing to help them.

4. Conflicts of interest

There are, then, cases where reproductive help may result in a child who is disadvantaged. The future child's interests can come into conflict with those of the person or couple wanting the child. Let us first consider the ethical issue in terms of parents whose child is likely to have a terrible life: perhaps the potential parents have a history of cruelty to children.

Where people have children without help, questions are rarely asked about how good they will be as parents. There is a general presumption that people able to have children are free to do so. Apart from a very few (and highly controversial) cases (such as sterilizing some mentally handicapped girls), society is reluctant to interfere with people's decisions to procreate, despite the risk of harm to the children. Normally, the most society does is to remove the child from parental care.

However, where the help of artificial techniques is sought, the question of what kind of home is best for the child comes into greater prominence. This is because the physicians and others whose help is sought may have some responsibility for the consequences of their assistance. In natural procreation, the presumption in favour of liberty creates a reluctance to raise the question of the qualifications of other people for parenthood. A doctor may feel that the liberty of others is one thing, but being asked to help bring about a possibly disastrous parenthood is another.

Because of the differences between respecting the liberty of others and helping them in their projects, to withhold help from some of the infertile need not be to start on a slippery slope towards compulsory sterilization of 'unsuitable' potential parents. Nor is it an invasion of bodily integrity in the way compulsory sterilization would be.

But a distinction has to be drawn between two very different types of case. If it is a question of giving a child to people who will behave with horrifying cruelty, it may be 'a mercy if the child is not born'. On the other hand, perhaps a child with only one parent suffers some disadvantage relative to other children, but few such children would feel life was anywhere near so bad as to wish they had not been born.

A parallel can be drawn with handicap. Some handicaps are so terrible that death can seem a mercy, and it is possible to think it would have been better if the person had not been born. But many handicaps are mild or only moderately severe. No-one would think that colour blindness made it better for someone not to have been born. No-one would think it wrong to help a couple have a child who would be at risk of colour blindness.

Just as there are degrees of handicap, so there are degrees of social disadvantage, and having only one parent perhaps comes towards the mild end of the spectrum. In the case of mild or only moderately severe social disadvantage, it may be misleading to say there is a conflict of interest between the future child and its potential parents. If the disadvantage is only relative, and the child will come nowhere near regretting having been born, it is hard to see how it can be in the interests of the future child for the potential parents to be denied help.

There is a question about who should decide whether help should be provided. Is it a decision to be left to couples and those they ask for help? Or should there be public policy, with regulations or guidelines? The issue is one of a conflict between liberty on the one hand and possible harm on the other. There should be a presumption against bureaucratic interference in these decisions about having children, but there can be a case for intervening to prevent children being harmed.

Moreover there is the point that being born with a mild disadvantage is not against the interests of the child, where the alternative is not being born at all. This, together with the difficulty of predicting how good a parent someone will be, suggests that, from the point of view of the child, only the likelihood of a fairly serious disadvantage supports guidelines ruling out help for potential parents.

5. The impact on the wider society

Such a concept as 'moderately severe social disadvantage', mentioned above, raises the possibility of these decisions about reproductive help having an impact on society at large, as well as on those directly affected. While the number of, say, single people seeking help to have children remains very small, any general social effect may seem too insignificant to consider. But suppose it became so widely accepted that a substantial proportion of births were to single parents. (There is no strong reason to think this likely, but the case is worth considering to get at an underlying theoretical issue.)

If such children were at some moderately severe disadvantage, and started to form a substantial part of the population, would this be worse than the present state of affairs?

It could be maintained that it would be worse, on grounds which can crudely be expressed by saying that the average level of happiness would be lower. ('Crudely', because talk of averages suggests a numerical measure of happiness we do not have.) But the principle that it is better if people who would reduce the average happiness are not born should be rejected. It has the consequence that all those in a population who are below the average level (in some cases perhaps something like half of them) are people whose existence should be regretted. It has many other absurd consequences, for instance that whether someone's birth is a good thing may depend on whether they are born into a country with a higher or lower average.

If there are some extra people whose lives are worthwhile, the fact that they have some relative disadvantage does not mean that their existence makes the world a worse place.

Yet there is a complication. The existence of extra, rather less happy, people does not *in itself* make the world a worse place. But there could in some cases be spill-over effects involving disadvantages for other people, which, if sufficiently magnified, could outweigh the case for bringing them into existence. There are very abstract arguments in the philosophy of population policy which show that accepting the case for 'extra, rather less happy people' can, when combined with other apparently innocuous policies, lead to disastrous results. (See Derek Parfit: *Reasons and Persons*, Oxford, 1984, Part Four.)

Here, these highly abstract arguments can be replaced by some more intuitive considerations. Suppose we could predict that the children suffering some moderately severe disadvantage were much more likely to become antisocial or violent. The existence of a very large number of them might have a dramatic effect on the crime rate and in other ways make society a worse place.

We can come closer to anxieties which have been expressed about reproductive help in some of these controversial cases. Could a substantial shift in the proportion of children born to single parents or to homosexual couples damage the institution of the family? Does society have an interest in preserving the traditional family? Is that in conflict with the desire not to discriminate against potential parents who are single or homosexual? These are illustrations of the way issues which at first sight looked mainly 'medical' turn out to raise far wider questions about the sort of society we want. We turn now to the impact on the institution of the family.

Chapter 4
The Family

Our ability to separate social from biological parenthood may create new patterns of relationship, such as families with two social parents of the same sex. Or one woman may be both aunt and mother to the same child. The family has for so long been the setting for our deepest relationships, both in childhood and in adult life, that any major changes need very careful thought. Even the possibility of changes may widely be seen as a threat. We believe that an unspoken general anxiety about the family often underlies the more specific disputes on the surface of the public debate on the new techniques of reproduction.

Sometimes this concern does come to the surface, as in the Catholic Church's document on these topics: *Instruction on Respect for Human Life in its Origin and on the Dignity of Procreation*. This document said of artificial fertilization using either semen or eggs from a donor that 'it brings about and manifests a rupture between genetic parenthood, gestational parenthood and responsibility for upbringing. Such damage to the personal relationships within the family has repercussions on civil society: what threatens the unity and stability of the family is a source of dissension, disorder and injustice in the whole of social life'. Non-Catholics, as well as Catholics, may feel concerned that artificial forms of reproduction perhaps threaten the unity and stability of the family.

These new forms of reproduction may change the family. But their development may itself have been caused in part by changed attitudes to the family. Why have these methods emerged now rather than thirty years ago? No doubt our more sophisticated science and technology are important. But social decisions (for instance about which projects of scientific or medical research to support) influence which knowledge grows and which technology is developed. Perhaps a more flexible view of the family has made us more open to accepting these kinds of medical intervention in reproduction.

1. The traditional family and some changes

Historical generalizations are usually simplifications, admitting of many exceptions. Aware of this, we tentatively offer a rough sketch of a traditional ideal of the family, perhaps with its roots in nineteenth-century Europe, and of some ways in which the family nowadays often differs from it.

The foundation of the traditional ideal family was a heterosexual couple embarking on a monogamous marriage for life. Adultery and prostitution existed, but were seen as a falling away from the ideal of the exclusive monogamous relationship. The rarity of contraception meant there were a lot of children. The rarity of divorce meant that the stability of the family could be taken for granted as the children grew up and as the couple grew old. The father earned the money and had authority in the family, while the mother cared for the children and looked after the home. (Nineteenth-century middle-class families more often came close to this ideal than did working-class families, where the woman often had to work.)

The spread of contraception has helped to change the picture. Families have fewer children, and mothers more often go out to work. This, together with other changes of attitude, has weakened the traditional authority of the father, who is no longer the unique provider. Contraception has also diminished one of the dangers of extramarital sex. The longer expectation of life may have added to the pressures on the traditional family. When people marry now, their partnership has a longer time in which to survive or break up than would probably have been the case a century ago. The spread of divorce has made children (and their parents) less able to rely on the permanence of their family.

Marital breakdown can be devastating for an abandoned spouse, and also for the children. Security and stability are so important for both children and adults to flourish that the traditional family is often and understandably seen as something to defend against further erosion.

But the drawbacks of the traditional arrangements are also often pointed out. The absence of divorce could imprison people in unhappy marriages. Couples had no choice about the number of their children. The size of a family was limited only by the brutal method of high infant mortality. Women had few opportunities for fulfilled lives except in the single traditional role of wife and mother. (Philip Larkin, in a poem about young mothers at the swing and the sandpit, says: 'something is pushing them to the side of their own lives'.) When the traditional family was at its height, social pressures meant that only the bravest looked for anything else. Men, tied to earning a living, were often effectively excluded from close relations with their children. Social pressures limited the children's role too. Parental domination over children could extend even to what job they took or whom they married. And a social structure based so much on the family left no recognized social place for homosexuals.

People of libertarian outlook, together with feminists and those who campaign for homosexual rights, see the loosening of the traditional family structure mainly as gain. Those more conservative about the family see the mounting divorce rate as a disaster. Disagreement about the relative gains and losses is not something this committee need try to resolve; we note that the changes have created a climate in which artificial reproductive techniques are more readily accepted.

More frequent marital breakdown has made it more common for children to be brought up by a single parent or by an unmarried couple. A more tolerant sexual outlook makes it less likely that the use of donors will be seen as a form of infidelity. The spread of contraception has enabled people to regard having children as a matter of choice, and also to think more readily of sex independent of procreation. As a result, it seems more natural to extend choice about children to the infertile and to accept procreation independent of sex.

Changing social attitudes have made assisted reproduction more readily accepted. But assisted reproduction may in turn be expected to shape our outlook. As these techniques become more common, there may be some pressure on people to use them in preference to natural conception. Some critics of the GIFT technique, where the fertilization takes place inside the woman's body, argue that IVF is better because it can be monitored. More fundamentally, the uses made of reproductive technology may change our conception of what it is to be a mother or a father.

2. The idea of parenthood

Parents create their children but, beyond this, they are not responsible for their biological characteristics. This may change if it becomes easy to choose the sex or other genetic characteristics of one's child. Not to make these choices will itself be a choice. And being a parent may come to be seen as creating one's children in a much fuller sense than now. The relationship with our children may be changed if they can criticize our choice of their characteristics.

Most of these problems are some way off yet. But our present use of donors and surrogates is starting to raise questions about what it is that makes someone a mother or father. There is a tendency to think of parenthood as an exclusively biological concept. But this view underrates the role of social convention in determining *which* biological links are to count as parenthood. This is brought out well by how we think of some of the different biological relationships in various kinds of donation or surrogacy.

In everyday (as opposed to legal) thinking, the man who provides the sperm in AID is not meant to be considered the father, whereas in surrogacy he is. Carrying an embryo provided by someone else makes you a mother in IVF but not in surrogacy. In other words, we pick and choose among the biological criteria because we want to designate the couple who intend to bring up the child as the parents, rather than the donor or the surrogate.

In all these cases, unlike that of adoption, the couple's claim to parenthood is backed up by at least one biological link, whether semen, eggs or pregnancy and childbirth. If donation and surrogacy become very common, we may come to see parenthood as primarily a social concept, and only secondarily a biological one. Perhaps this shift to a less biological perception of parenthood will in turn affect how we see our bodies and our sexuality.

Our concern about tampering with the family and parenthood has a lot to do with the great importance for a child of the warm and stable relationships a family can provide. This reason for caring about the family is not necessarily a reason for opposing all changes in its nature, or in the concept of the family. There may be gains if we adopt a more flexible concept of parenthood, for instance allowing the possibility of double fatherhood, rather than thinking that either the social or the biological father must be *the* father. (Something similar may apply to children whose parents divorce and remarry.) On the other hand, if a fundamental part of the father's role is to take part in important decisions affecting the child, it might be confusing for the child to have more than one father. An alternative would be for the donor to be seen as like an uncle or a godfather: someone having a special link to the child, but not necessarily involved in his or her upbringing.

We do not have to think of the family in rigid all-or-none terms, such that families deviating from the standard type somehow do not count as real families. (A recent law in Britain contains a clause preventing local authorities from spending public money on any literature portraying homosexual couples with children as a 'pretended family'. On a less rigid view of the concept of the family, there can be room for evolution. Few nowadays are likely to feel that it would have been better when divorce became common to insist that divorced people who re-married were engaged in 'pretence'.)

To suggest acceptance of the idea of double fatherhood is not to say that it would in all cases be good if the donor were thought of by the child as a second father. In these relationships, there is no policy which can mechanically be applied. In all these issues about donors and parents, no set of rules is likely to be flexible enough to fit the subtleties of different people and different relationships. What is best for one child in one family may *not* be best for another.

Conceptual changes are likely to be gradual. But decisions about links with donors will have a more immediate impact on the kind of family in which the children grow up. So will those about who should be helped to have children.

3. Access to help in having children

Some people think that reproductive technology should only be available in the 'standard' case of the infertile couple. Others disagree. A doctor who wants to stick to helping the infertile couple should of course be free to follow his or her conscience. But what should we say about a doctor or clinic deciding to give help in the non-standard cases? Are they doing good by helping satisfy the need for children in women who would previously have been denied them? Or are they doing something to which there are ethical objections?

The non-standard cases are not all the same, and it may be consistent to take one view of the widow and a different one of the lesbian couple. We will here briefly take the case of the lesbian couple as an example. This is an interesting case for its possible impact on how the family is to evolve.

Semen donors may not want their biological children to grow up with two lesbian parents. But this is not a central issue, as no doubt some donors could be found. (Apart from sympathetic heterosexuals, some male homosexuals might see donation as a contribution to defeating prejudice.) The interesting ethical issues start when this problem is overcome.

What is the case *against* helping? In these non-standard cases we would not be overcoming a medical problem, but would be circumventing biological limits to parenthood. The normal state for a child is to have one parent of each sex. It is surely right to be very cautious about tampering with something so fundamental, involving the possible risk of psychological damage. For instance, there is the question of how the child will think of sexuality and procreation. And growing up in a lesbian family might frustrate a boy's need for a male role model, or a girl's need to develop an intuitive understanding of the other sex. (The same points could be made in reverse about children being brought up only by men.) If the lesbian couple becomes a common alternative nucleus for a family, any psychological losses will become accordingly widespread.

What is the case *for* helping? Homosexuals have suffered, and still suffer, from an appalling degree of prejudice and discrimination. This has often made homosexuality a sad condition, and the extension of the idea of the family to include a version with lesbian parents is a large move towards greater equality.

It could be said that there cannot be equality *in procreation* between heterosexuals and homosexuals. Just as the heterosexual woman who does not want children must always take this possibility into account (for instance by using contraception), so must the lesbian who wants children accept that she cannot have them with her female partner.

One reply to this is: just as heterosexuals use contraception, why should not lesbians use reproductive technology? Such a reply is in one way an over-simplification. There is a difference between using contraception to control one's own reproductive capacities and using reproductive technology. The latter helps homosexuals by introducing heterosexuality into the relationship in a depersonalized form, via the semen donor. There are complications about the status of the donor in a lesbian family. It would be possible to admit of lesbians having children through adoption, while wishing to draw the line at tampering with the idea that reproduction requires people of both sexes.

Yet it is surely right to be predisposed in favour of anything that removes some of the barriers against homosexuals having a fulfilled family life. Lesbians who want to have children are not different in their needs from heterosexual women. Like many other women, lesbians may care about what adoption does not provide: having a child genetically theirs and to whom they give birth. They may care as much about having *their* children as an infertile wife, and their lives may be as much enriched by such children as anyone else's.

4. Some thoughts on policy

Removal of discrimination against homosexuals in such a fundamental matter would be a great gain. But obviously there are strong grounds for unease about reproductive help where the family circumstances may impose a serious handicap on the child. The anxieties are based partly on the child's own interests, and partly on the social impact of the spread of such families. Everything depends on how well founded these anxieties are. People differ over the hypothesis that children born through AID to one of a lesbian couple are likely to be at a disadvantage. Some think that, in the light of what we know of human nature, the hypothesis is very plausible, while others think there is no such presumption in favour of it.

The fullest study we have been able to find is suggestive, but not conclusive (Golombok, Spencer and Rutter: *Journal of Child Psychology and Psychiatry*, 1983).

Thirty-seven children aged between five and seventeen, being brought up in twenty-seven lesbian households, were compared with thirty-eight children in twenty-seven single-parent families, being brought up by a heterosexual mother. Psychosexual and psychiatric appraisals were based on interviews with the children, on interviews with the mothers, and on questionnaires given to the mothers and to teachers. The two groups did not differ in gender identity: all the children said they were glad to be the sex they were. (But gender identity is usually established at an early age, and in some cases *could* have been established before the lesbian partnership was set up.) The two groups did not differ in sex-role behaviour. And there were no signs of differences in sexual orientation between the two groups. (This was based on the reported patterns of friendship in pre-pubertal children, and also on reports of romantic friendships in the small number who were past the age of puberty.)

These negative findings suggest that anxieties about the effects of being brought up in a lesbian family may be unfounded. But they cannot completely exclude the hypothesis of disadvantage.

Because the control group were in single-parent families, conclusions depend on the evidence that children brought up by single parents are not themselves different in these respects. To test the disadvantage hypothesis, it would be more satisfactory to have a direct comparison with children brought up by heterosexual couples. And the great majority of the lesbians in the study had previously had heterosexual relationships. (Of those who had not, one had an adopted child, and the other had one adopted child and one child by AID.) It *may* be misleading to extrapolate from that group to all those seeking reproductive help, if the majority of the latter group have not had heterosexual relationships. Also, as the authors of this study themselves stress, the question of sexual orientation cannot fully be settled without a follow-up study when the children are older.

If children were worse off having two female social parents, this would have to be set against the elimination of discrimination against homosexuals. But the claimed disadvantage, while not finally refuted by the evidence, is certainly not supported by it. And, if there is any disadvantage, it is hardly likely to be so bad that it would be better if the children had not been born. So the objection is at most one about 'moderately severe' disadvantage.

We expect that, until the disadvantage hypothesis is overwhelmingly supported or refuted empirically, those who have to decide about helping will divide in their views, as we do. It seems right for doctors and others to follow their consciences, some giving help and some refusing it. (However, in some countries, such as Germany, professional guidelines forbid non-standard use of AID.) The result of those who believe in giving help doing so can be expected to be some growth in the numbers of such families.

This is likely to go further as feminist or lesbian groups start themselves to provide the help denied by more traditional clinics. In the United States, feminist health centres have started to do this; in 1982, the Oakland Feminist Women's Health Center set up the first feminist sperm bank (Francie Hornstein: 'Children by Donor Insemination: A New Choice For Lesbians', in Rita Arditti *et al: Test-Tube Women*, London, 1984). It is likely that self-help arrangements, supplemented by women's clinics where necessary, will spread in Europe.

The other possible basis for opposing reproductive help to lesbians is its impact on society. There are two parts to this case. One is linked to the hypothesis that the children are disadvantaged, and is based on the view that larger numbers of disadvantaged people may affect society for the worse. The other part of the case is the need to protect the institution of the family.

We take the view that the restriction of liberty involved in any legal ban on reproductive help to lesbian couples could only be justified by grounds for believing that harm would be done by that help. Because the view that the children would be worse off than others is itself speculative, losses to others as a result of their disadvantages are even more speculative.

It could be argued that future people would suffer psychological losses through the dilution of the traditional concept of the family. But, because this is again such a speculative claim, we think this case is insufficient to justify a legal ban. And we are sceptical about using the law in an attempt to freeze our changing family structure. One of the costs of widespread adultery is its weakening of that structure, but to criminalize it would seem a heavy-handed response. And that seems equally true of the more speculative weakening resulting from helping lesbians to be mothers.

While the members of this committee are divided about the desirability of providing help, we are agreed that it should not be legally prevented. Those of us who are inclined to think it wrong to help do not wish to criminalize the behaviour of those who take a different view. The case that the children are disadvantaged is too weak to support that.

Rather little is known with certainty of the effects of lesbian parenthood. On the one hand, this is an argument for caution. But, on the other hand, it can be an argument for letting the future shape of the family evolve experimentally. No doubt people should be discouraged from taking high risks of major family disasters. And it goes without saying that new forms of family life must only be tried voluntarily. But, subject to these qualifications, we prefer a society predisposed in favour of 'experiments in living' to one in which they are stifled.

We may find that *not* all happy families are alike.

Part Three
Surrogate Motherhood

Chapter 5
Surrogacy in Practice

Is surrogate motherhood something to welcome or to discourage? We look first at what it is like in practice, and then turn to the ethical issues.

Two different things can be meant by 'surrogate mother'. In one version, the surrogate mother's own egg is fertilized to produce the child she bears for someone else. In the other, 'womb leasing' version, the egg is fertilized *in vitro*, and is not hers. It may come from the woman who wants the child.

The term 'surrogate mother' implies that one woman replaces another in her role as mother. In the 'womb leasing' version, the replacement is that of the body for the period of pregnancy. In the other version, the surrogate is a replacement in two further ways. There is a symbolic sexual replacement: she is inseminated with the sperm of the other woman's husband. (She replaces the wife, and she abstains with her own normal partner, to avoid the risk of his being the biological father.) And, through her egg, the surrogate is the genetic mother.

Although the 'womb leasing' version makes the business more medical, it is said to be easier for the surrogate. It reduces the symbolic sexual replacement. And, as the egg is not hers, it reduces the feeling of giving away her own child.

1. Motives of surrogate mothers

A woman may, as an act of generosity, bear a child for her sister who is unable to do so. At the other end of the continuum, the arrangement may be a commercial one between strangers. It is hard to find surrogate mothers, so there is very little physical or psychological screening of them.

There is some evidence from the United States of the motives of both surrogate mothers and potential ones (Philip Parker: 'Motivation of Surrogate Mothers: Initial Findings', *American Journal of Psychiatry*, vol 140, no. 1, (1983), pages 117–118). Payment seems a necessary motive, but is not the whole story, particularly as pregnancy progresses. Some women enjoy being pregnant. Others hope to compensate for some past birth-related loss, such as an abortion, having given up a child for adoption, or having been herself an adopted child. A key element seems to be a longing for a close friendship with the future parents of the child. Other, French and English, cases fit this picture. The English surrogate mother 'Kirsty Stevens' wrote: 'I realized that if I ever did this, if I had a baby for another woman, the emotional closeness would have to be the most important aspect of the deed. I would need to get close to the people I had the baby for, so that in the end I would be giving them their baby as a gesture of friendship'. She says she turned down an offer of about sixty thousand pounds, choosing instead a couple who could afford ten thousand pounds but with whom she felt she could have a close friendship (Kirsty Stevens: *Surrogate Mother, One Woman's Story*, London, 1985).

2. Payment and the contract

The surrogate and the couple often come together through an advertisement, issued by an intermediary; usually legal in the United States and medical in such European countries as France and Denmark. There is an explicit or implicit offer of a fee, which makes response more likely from women in financial need. The couple paying the surrogate will normally be richer, especially in the United States, where the lawyers also have to be paid. The poorer surrogate may be open to exploitation.

The payment may be seen as a salary for work done, or as compensation for the expenses involved in pregnancy, or else as a gift intended to reciprocate the surrogate's generosity. The couple paying the surrogate normally do not want a continuing relationship with her. The payment is meant to close off any debt and to end the relationship. Sometimes this is underlined by a contract.

In some countries, such as France and Germany, a surrogacy contract would not be enforceable by either party. In the United Kingdom, the Surrogacy Arrangements Act makes commercial surrogacy a criminal offence. In some states in the United States, on the other hand, surrogacy contracts may be legal, although it is not certain that they are enforceable.

The couple want the contract to protect themselves and the child. They want the surrogate to hand over the child and end the relationship. They also want the surrogate to do whatever is necessary while pregnant to reduce the risk of the child being handicapped.

Because the couple want to end the relationship, often the surrogate is not given their name and address. Anonymity is usually not complete. The couple may contact the surrogate during pregnancy, and are sometimes present at the birth. Noel Keane's agencies in the United States do not encourage total anonymity. (It should be mentioned here that commercial surrogacy is now illegal in Michigan, where Noel Keane had his biggest practice.) Both surrogates and couples interview several alternatives, so that they can choose who they would be most comfortable with. But the agreement almost always explicitly stipulates that the surrogate will not later try to locate or contact the child. The couple are anxious to exclude both her and the questions about kinship her presence might raise. (In several European countries, unless she gives up the child for adoption, from a legal point of view she *is* the mother.)

3. Pregnancy and birth

The couple want a healthy child, who is unambiguously fathered by the husband. As a result, the contract may impose conditions on the surrogate which arouse disquiet about exploitation.

Intercourse with any other partner around the time of the conception will be excluded. A blood test may be done on the child at birth to check that it was conceived with the sperm of the right man. Medical examinations may be required during pregnancy. Amniocentesis and other tests for foetal abnormality may also be required. The surrogate may be expected to have an abortion if a test is positive. She may also be expected to stop smoking or drinking. Because the child is explicitly defined as not hers, the surrogate may not dispose of herself or the child during pregnancy as she sees fit. Here the interests of the potential parents and of the future child conflict with the freedom of the surrogate mother to make her own decisions. Although she may accept that the child is not hers, these decisions are still about her body and about her pregnancy and childbirth.

4. After the child is born

Even when things go according to plan, there can be a problem which is often overlooked. The natural children of the surrogate may be disturbed by what may look like her giving away one of her own children.

There are two ways in which the surrogacy arrangement can go wrong after the baby is born. The baby may not be accepted by the intended social parents: perhaps the intended social mother has died or developed a serious disability, or the couple refuse to accept a baby born handicapped. Or the surrogate mother may change her mind and want to keep the baby.

Where the couple do not accept the baby, there are similarities with cases where a baby is rejected by its natural parents; but there is a special problem for the surrogate mother, who may be appalled by the abandonment of the baby she has given birth to, and who may as a result feel under pressure to look after a child she did not intend to keep.

When the surrogate mother wants to keep the baby there is a particularly poignant conflict. Her attitude towards the baby may change completely during pregnancy. Confidence in advance about being willing to give up the baby is not a reliable guide. And in some countries (e.g. Denmark) she cannot formally give up the child for adoption until three months after the birth. If she does keep the baby, this shatters the expectations of the couple who were expecting the child, and leaves the father feeling that *his* child has been taken away.

But, if the contract is enforced, the surrogate mother can feel shattered. When, at one stage of the 'Baby M' case in the United States, the judge decided against Mary Beth Whitehead, a previous surrogate mother (Elizabeth Kane, who gave up her son in 1960) expressed this feeling: 'I was just devastated by the decision. I feel as if somebody died. When I heard the news that Mary Beth had lost all visitation rights, I suddenly thought of how my son looked the last time I saw him, in his cradle when he was two days old, and I thought of all the Christmases and all his birthdays without him, and all the grief I've gone through, and I thought, "Oh my God, Mary Beth is going to have to go through all that too". I was just crushed.' (The decision was later reversed by the New Jersey Supreme Court. Commercial surrogate mother contracts are now unenforceable in New Jersey.)

5. The issue of exploitation

In the 'Baby M' case, the contract had been drawn up by the Infertility Center of New York. The three parties were the Center itself, the surrogate mother Mary Beth Whitehead, and the natural father William Stern. The case was not one of mere 'womb leasing': Mary Beth Whitehead was as much the genetic mother as William Stern was the genetic father.

The Center was to be paid seven thousand, five hundred dollars by William Stern, in advance and non-refundable. It was to be exempt from all guarantees either that Mary Beth Whitehead would become pregnant, or that she would surrender the baby. Mary Beth Whitehead was to receive ten thousand dollars, but only after the surrogacy was over. She contracted to 'assume all risks, including the risk of death'. The contract would be over, with no compensation to her, if there were a miscarriage in the first five months. If tests showed abnormalities, she was to have an abortion 'on demand of William Stern'.

Mary Beth Whitehead eventually settled her claim for damages against the Infertility Center of New York. But perhaps a comment made at an earlier stage still stands. The journalist Murray Kempton wrote that 'the only party to this contract to take account of the errancies of human sentiment was the Infertility Center of New York, and it insulated itself against them with a calculation so cold as to embarrass a social order that licenses as a service works like these'. (New York Review of Books, 9 April 1987.)

Chapter 6
The Ethics of Surrogacy

The two central questions are whether surrogate motherhood is morally acceptable and whether it should be legally permitted. These questions are not identical, but they are related.

If any policy short of a total ban is adopted, the main further questions are these:

Should there be surrogacy contracts? Should they be legally enforceable either when the surrogate breaks the restrictions imposed on her during pregnancy, or when she changes her mind about handing over the child?

Should the contents of contracts be regulated, either to protect the surrogate from exploitation, or to protect the interests of the child?

Should there be screening of surrogate mothers? Possible problem cases include heavy smokers and people with severe psychological problems. They also include the sister or other close relation of the potential parent, who may be particularly likely to volunteer, but whose role might create family problems later.

What is the role of intermediaries, such as doctors, clinics and agencies? Should there be profit-making agencies?

1. The case for surrogacy

Part of the case is straightforward. Surrogacy relieves childlessness. For women who have had repeated miscarriages, or who suffer from conditions making pregnancy dangerous, surrogacy may be the only hope of having a child.

Another, more problematic, argument appeals to the interests of the child who would not have existed without surrogacy.

Another argument appeals to liberty. Some strong justification is needed for preventing people from bearing children to help their sisters or friends. And a similar strong justification is needed for preventing people freely contracting to do this for someone for money. This argument relates to the legality of surrogacy, but does nothing to show that surrogacy is a good thing in itself. The central case for that has to rest on relieving the burden of childlessness.

2. The case against surrogacy

(a) The children: One line of thought appeals to the rights of the child. It appears in the Catholic document issued by the Congregation for the Doctrine of the Faith, which says that surrogacy 'offends the dignity and the right of the child to be conceived, carried in the womb, brought into the world and brought up by his own parents'.

This case seems to us not overwhelming. Even if the child has a strong interest in being created sexually, to call this a *right* is to claim that it trumps *any* interests of the childless couple. This requires that being the child of a surrogate is such an indignity that, by comparison, relieving *any* degree of the potential parents' misery is to count for nothing. We have not found the powerful supporting argument this would need.

The objection is made even weaker by a further problem. For the potential child, the alternative to surrogacy may be non-existence. It seems unlikely that the child will see surrogacy as so bad as to wish he or she had not been born at all. The 'right' looks like one the child will later be glad was not respected. It is hard to see the case for giving this supposed interest any weight at all, let alone for saying that it justifies leaving people unwillingly childless.

Another argument appeals to the psychological effects of surrogacy on the child. If the surrogacy is paid for there is a danger that the child will think he or she has been bought. Also, it is sometimes suggested that surrogacy breaks a bond formed by the time of birth. Dr John Marks, the chairman of the British Medical Association, has said: 'By the time a baby is born there is a bond between the mother and the child. With surrogacy you break that bond. You are depriving the child of one natural parent. We think that is wrong' (The *Guardian*, May 8, 1987). It is reported that the General Medical Council may ban doctors from involvement in surrogacy. This step has already been taken in West Germany.

The surrogate mother may well feel a bond between herself and the child. But is there reason to believe in any bond in the other direction before birth? Or could this be an illusion created by projecting the mother's feelings on to the foetus? If the child's feelings are a reason against surrogacy, the baby has to have, by the time of birth, highly specific feelings towards the particular woman who bears him. The evidence for this can charitably be described as slight.

Suppose, for the sake of argument, that there is such a bond. It is then undesirable to break it. But, where it is broken, is the child so harmed that it would have been better if he or she had not been born? For this is what banning surrogacy on these grounds seems to imply. We do not have such drastic thoughts about people who are adopted. The British Medical Association's Board of Science is quoted as saying that while adoption may be 'the next best thing' for a child facing an uncertain future, any arrangement where a surrogate mother hands over the child 'dooms it to second best from the start' (*The Independent*, May 8, 1987). But is it obvious here that no life at all is preferable to 'second best'?

(b) Conflicts: The conflicts sometimes arising between the potential parents and the surrogate mother may harm the child, and this is part of the case against surrogacy.

(c) Effects on the family: Perhaps introducing a third party so intimately into the process of having children may weaken the institution of the family. (In West Germany, the report of the Benda Commission considers a legal ban on surrogacy with the exception of surrogacy by relatives.)
(d) The surrogate mother: The position of the surrogate mother varies, according to whether she is bearing a child to help a sister or friend, or has made a commercial arrangement. There is the criticism that surrogacy is an invasion of her bodily integrity. This criticism may be weaker if she willingly agreed than if she was forced into it by money problems. Sometimes she may bitterly regret having agreed to give away the baby. As we have seen, there is a danger of her being exploited. Financial pressures may put her in a weak position to resist contractual conditions which give little weight to her interests.

Another important motive for volunteering to act as a surrogate seems to be the desire for friendship with the parents-to-be. As this is usually exactly what the parents-to-be do *not* want, it is an illusory objective. She wants friendship: she is treated as a provider of a service, and afterwards dismissed.

3. Policy

Is surrogacy something to encourage or not? The Warnock Committee said, 'The question of surrogacy presented us with some of the most difficult problems we encountered'. We found this too. The central issue is the conflict between the interests of the childless couple and those of the surrogate mother.

Some members of the committee are opposed to surrogacy in principle, because of what the practice does to the surrogate mother. The invasion of her bodily integrity, the disappointment of any hopes for friendship with the family who receive the child, the psychological trauma of giving up the baby, and the possibility of regrets for the rest of her life, add up to a very strong case against surrogacy. There is also the possibility of ill effects on the surrogate's own family, particularly on her own children.

Other members of the committee share these anxieties, but

are sufficiently impressed by the needs of infertile couples to think that some cases of surrogacy are beneficial. Whichever view we take, we agree both that surrogacy should not be illegal, and also that, if it *does* take place, it should be subject to certain restrictions.

A general legal ban would be unenforceable. Moreover there is a powerful consideration which also influenced the Warnock Committee: the birth of a child should not have a taint of criminality. We recommend that surrogacy should not itself be illegal.

The surrogate mother is notably vulnerable, and any acceptable arrangements for surrogacy must give her a lot of protection. The child should be protected against prolonged battles between the surrogate and the potential parents. These two considerations support restrictions which may in practice greatly reduce the frequency of surrogacy. While this will leave some infertile couples childless, we think this is a lesser evil than the ones such restrictions would be designed to avert.

4. Agencies and regulation

Sometimes surrogates will be relations or friends helping people they know well. But there is a case for the option of making arrangements through a clinic or other agency. They will have more experience of the problems than the couple or the surrogate mother. They will know what should go into a contract. And they will be able to carry out any necessary screening either of couples or of surrogate mothers.

Where these agencies are public, they should operate on guidelines open to public inspection. If there are private agencies, they should be publicly inspected and licensed.

5. Making contracts unenforceable against the surrogate mother

We think that the surrogate mother should remain free to decide for herself whether or not to have an abortion. And we think she should not be forced to hand over the child against her will. In these respects at least, any contract should not be enforceable against her.

This is mainly to protect the surrogate mother. But there are other reasons. If the contract were enforceable, a pregnant surrogate who started to change her mind might become depressed. This could be bad for the child. If a contract were enforced against a reluctant surrogate the couple might feel guilty, which could interfere with their relationship with the child. If the child found out (as would be likely in a system not based on anonymity), the relationship might again be disturbed.

6. The claims of the biological father

This policy of unenforceability supports the claims of the surrogate mother against those of the biological father. In cases of dispute over the baby, some think that the father's right to his child should not be overridden. There are two main arguments for this: he has a right arising out of a contract freely entered into by the surrogate, and he *is* the biological father.

(a) The contract: This is partly a claim about a legal right. In many countries, such contracts are unenforceable, and so this legal right does not exist. The issue is whether they *should* be enforceable, and this cannot be settled by citing what the law happens to be.

Most legal systems refuse to recognize some kinds of contract, such as those in which people sell themselves into slavery. This refusal is a restriction of liberty: it excludes people's freedom to bind themselves in certain ways. It is a protection against people doing themselves great harm or giving away vital liberties. It is a form of 'paternalism': restricting people's freedom in their own interests.

Some are opposed to all paternalism. Too much paternalism can be a danger to liberty. But the case of contracting to become a slave illustrates the difficulty of excluding all paternalism. Stopping people making enforceable slavery contracts protects one of their vital interests. It also protects far more freedom than it takes away. We think that these reasons (as well as the other non-paternalist ones) apply to making surrogacy contracts unenforceable.

(b) The biological link: The other basis for the father's claim to take the child from the reluctant surrogate mother is that he *is* the biological father. This case also seems to us not decisive. The father is genetically linked to the child. But so, in one form of surrogacy, is the surrogate mother. And even in the other form, 'womb leasing', she has still carried and given birth to the child. This too is a biological link, and often creates, on her side at least, an emotional link as well.

The father's interests are not negligible here. Because of the genetic link, he is likely to care a lot about the child. And, if the surrogate changes her mind, he and his partner will have the anguish of childlessness compounded by disappointed expectations.

But it is not obvious that, of the biological bonds, the genetic one should trump the others to give the father a right to the child. And, severe as the couple's disappointment is, we do not think it justifies forcing a woman to endure the anguish of being made to give up the child she has given birth to.

The possibility of the surrogate mother changing her mind should be accepted by the couple as one of the risks of this way of trying to overcome childlessness when embarking on it.

7. Protecting the child

If there is to be surrogate motherhood, the child needs protection from emotional damage. In every way things should be as normal for the child as possible. This is one reason why surrogacy should not be illegal. It cannot be good for children to know that their social parents or surrogate mother committed a crime through having them.

It is also surely bad for children if there are long legal battles over them. And even if the battles lasted only for a few months, the insecurity in the (social or surrogate) mother looking after the child could interfere with bonding, as would any transfer which then took place. These are further reasons for making the contract unenforceable in this respect against the surrogate mother. Her decision to keep the baby should be final.

But they are equally reasons for making her decision *not* to keep the baby final. Once the baby has been voluntarily handed over to the social parents, further upheavals around the child should be avoided. The social parents should be recognized as the legal parents beyond further challenge, even if the surrogate later changes her mind.

The surrogate mother should normally be regarded as having no right to a further relationship with the child who has been handed over. Visits or other contact could undermine the child's security about who his or her parents are. There are special cases, as when a relation or close friend acts as a surrogate. But, in other cases, we think that the child's interests require severance of the relationship to be the norm. (Though there is a case for the child having the right to be told on reaching maturity the identity of the surrogate mother. The case is like that for non-anonymity of donors already discussed.)

8. Screening potential social parents

No individual should be put in the position of breaking the law by becoming a surrogate or by contracting with one. But agencies could adopt a policy of screening potential surrogates or potential parents, and it would be possible for *them* to be legally required to do so.

We think there should be a background presumption that reproductive help should be available to those who request it, subject to competing claims for resources, and with certain exceptions where special reasons apply. But, because of the special problems of surrogacy, particularly for the surrogate mother, we think that agencies should only help those who cannot have children in other ways, or where serious medical risks are involved. The desire not to interrupt a career does not seem sufficient reason for imposing the risks and traumas of surrogacy on another woman, even if she is prepared to accept them.

9. Screening potential surrogate mothers

There is now little screening of surrogate mothers. They are not easily found, and screening may seem just a way of reducing their number still further. But some screening, or at least counselling, would be in the interests of everyone. It is possible, for instance, that women who have been very promiscuous, or who have had a long series of unstable relationships, or who have been abandoned in childhood, are more likely to resist giving up the baby, or else to suffer from extreme depression over it.

Some screening would be on behalf of the potential surrogate herself. Apart from the psychological aspects, she may want to know that pregnancy will not carry any special health risks for her. The screening period might give her time to think about her decision, and about its implications for herself and for her family. The agency should provide her with counselling help during this period.

Some screening should be on behalf of the child. Heavy smokers, alcoholics and other addicts may harm the child and these conditions should be grounds for exclusion. (*Grounds* for exclusion: where surrogates are very hard to obtain, these grounds may *perhaps* be overriden. From the later perspective of the child, it may be better to have run those risks, or even to have suffered some harm, than not to have been born at all.)

Some screening should be on behalf of the potential parents. The kind of history linked with the surrogate changing her mind about handing over the baby seems a reasonable ground for exclusion. There are other grounds. In one case, a surrogate repeatedly extracted more money from the potential parents, threatening to kill herself unless they paid (Noel Keane and Denis Breo: *The Surrogate Mother*, New York, 1981, quoted in Peter Singer and Deane Wells: *The Reproduction Revolution*, Oxford, 1984, pages 116–117). No doubt it is hard to pick up this sort of thing in advance. But, where there is reason to suspect it, the potential parents would have reason to complain if the agency accepted the woman as a surrogate.

There are obviously difficulties in predicting who will be a good surrogate mother. Those rejected by agencies would be free to be surrogates by private arrangement. These comments about screening are intended as general guidance. Good agencies will revise their criteria in the light of their experience, and in the light of evidence collected in studies of other cases.

Chapter 7
Having Children and the Market Economy

With both surrogate mothers and semen donors, the question of payment comes up. We discuss these issues briefly, and then turn to the more important issue of whether there should be commercial agencies for surrogacy.

These may be straightforward questions about efficiency, to be answered by seeing what works best. In part, they are questions of that sort. But, entangled with those pragmatic problems are some deeper issues about what sort of society we want.

1. Semen donors

We have seen that some hospitals and sperm banks pay semen donors, while egg donors are rarely paid. Is it better when semen donors are also unpaid?

The case for payment is mainly that it brings in more donors, as the Necker Hospital found when it stopped paying. Payment also gives some recognition to the donor. And it is a way of ending the donor's involvement.

On the other hand, as with blood donation, payment may lead unsuitable people to apply, lying about their medical history (perhaps one of Aids) in order to be paid (Richard Titmuss: *The Gift Relationship, From Human Blood to Social*

Policy, London, 1970, chapter eight). And, although Necker Hospital found that unpaid donors were fewer, they were more diverse in their socio-economic background.

There are further arguments against payment. By giving donors a 'reason' to account for their donation, it may incline them to think less about the implications of what they are doing. Is our aim to obtain the maximum number of donors, no matter how, or do we want men to think about what they are going to do before they do it? Semen donation may mean nothing to the traditional anonymous medical student, but he can be motivated by payment. On the other hand, a man who has himself either experienced difficulty in having children or who has had an infertile couple in his family or among his friends, may be someone to whom donation makes very good sense. He may find reasons for giving semen without having to be motivated by money.

Payment also deprives donors of the chance of doing something purely for others. Blood donors sometimes say that paying for blood would debase the value of the act, and make it feel less worth doing. Payment and non-payment seem to lead to two different conceptions of donation.

Our inclinations are strongly towards a non-commercial ethos for donation, and we are impressed by the ethos of some of the systems developed in France. An economic case is always easy to understand, as is a case based on a quantifiable change, such as a larger or smaller number of donors. But policies adopted for such reasons can have more subtle side-effects, often of a kind not easily measurable. The issues we are concerned with are not only about easily measurable effects, such as how many infertile couples are enabled to have children. What matters most is how these techniques can be used to enrich people and their relationships, rather than diminish them. The *central* focus of this report is not technology but people. And so we think it right to stress the way payment affects the psychology of donation, turning what could be an enriching act of altruism into an act more like selling an old motor-bike. Every time we institutionalize the commercial solution rather than the altruistic one, we take a small step further towards a society where more relationships are permeated by the motive of economic gain.

Our preference for a non-commercial ethos is strong, both for semen donation and for egg donation. But we do not think the case against commercialism is powerful enough to justify a legal ban, which would anyway probably be unenforceable. We would like to see strong public campaigns for altruistic donation, bringing the plight of the infertile to the front of public attention. Only after such a policy has been tried should the market be thought of as a last resort policy to fill any remaining gap.

2. Should surrogate mothers be paid?

Surrogate mothers do something really important for the potential parents. Unlike semen donation, surrogacy involves a lot of inconvenience and some risk. Apart from women helping sisters or friends, unpaid surrogates are hard to find.

If surrogacy is to be available to childless people without altruistic sisters, payment may be hard to escape. There is again the question of whether we want to encourage the spread of the market into this area. Payment brings the danger of poor women being exploited. They may be pressured into surrogacy by their need of money. One way of making it less exploitative (apart from making the contract unenforceable, and excluding people who have health risks) would be to make the payment substantial. But, on the other hand, high pay may make it even harder for a woman needing money to resist.

In one way, the higher the pay the less the exploitation. But commercialism, with its pressure on poor women to embark on surrogacy they may later regret, seems a greater evil. It is better for fees to surrogates to be nonexistent or kept to a minimum. This may reduce drastically the number of surrogate mothers. That seems to us more acceptable than the commercial alternative. The payment of surrogates should not be illegal, as such a law would be unenforceable, and because of the need to avoid a child's birth being tainted with criminality.

3. Commercial agencies

The payment of sperm donors or of surrogate mothers at all lets market forces into childbearing. But it does so in a small way. Commercial agencies create much more of the ethos of the market.

Some of the arguments against market forces appeal to the public interest. For instance monetary incentives may encourage medically unsuitable people to sell blood or semen. And, in the present context, where would-be parents have paid a surrogacy agency, they may be more willing to engage in lengthy legal battles, even at considerable psychological cost to the child.

There is a case against banning commercial surrogacy agencies. A ban restricts people's freedom to earn their living in the way they choose. It also restricts the freedom of others by making surrogacy less available. If the agencies meet a demand, making them illegal will frustrate it. If there is no demand, making them illegal may be said to be unnecessary, on the grounds that they will go out of business anyway.

The suggestion that, if there is no demand for their services, the agencies will go out of business is naive. Commercial organizations can often create a demand for previously unwanted products. (An issue then arises as to whether such supplier-induced demands should be taken less seriously than others.) But the central argument is the appeal to liberty. It is true that banning commercial agencies is a restriction of liberty. But so is any law. And any law is either unnecessary or else stops some people doing what they want to do. There is a presumption against restricting liberty. Are there reasons powerful enough to justify doing so in this case?

A major argument for banning commercial agencies is to protect surrogate mothers from being exploited. But, desirable as this is, it might be done by regulating how agencies operate. Regulation is still an interference with the market, but one easy to justify on principles similar to those used to prevent companies operating factories that are unhealthy or unsafe to work in.

Many who oppose the extension of the market into this area would not be satisfied with commercial agencies being subject to controls to prevent exploitation. The opposition is not based only on worries about exploitation, but also stems from the idea that childbearing is simply not something to which buying and selling are appropriate.

There are two main arguments for limiting the market. One is based on equal access to certain basic goods. The other is linked to a preference for a society not dominated by money values.

(a) Equal access to basic goods: It is widely accepted that the services of the police should not depend on the citizen's ability to pay. Some of the arguments for non-commercial health systems have the same basis. Basic medical care and protection against crime are held to be such fundamental interests that they should be equally available to rich and poor. One argument against commercializing the new reproductive techniques would place help necessary for having children in this category of fundamental interests.

(b) Limiting the dominance of money values: Money is only a means of exchange for goods and services. But the dominance of money transactions can affect the values of a society, by making people less inclined to unpaid acts of altruism. And a society can be dominated by money in a different way: wealth or income can increasingly colour people's relationships.

4. The exclusion of commercial agencies

The case for equal access to basic goods carries some weight against commercial surrogacy agencies. But it could be questioned whether this kind of help with having children qualifies as one of the basic goods where money should be irrelevant. It would be perfectly consistent to think that essential medical treatment should be available equally, but to deny that this held for surrogacy.

The stronger argument seems to be that based on resisting the encroachments of commerce on the intimate relationships of parenthood. Even here the case can be questioned. If commercial agencies are banned, surrogacy will often be arranged through a non-commercial agency. And the intimate relationships to be protected are between man and woman, parents and children, not those with the non-commercial agency. So why should a commercial agency make any difference?

The difference is a matter of the way commercialism changes the way we see the things that are bought and sold. Almost inevitably they come to be seen as commodities. And some aspects of life seem particularly inappropriate to the market. Having children may be seen in this way because it is such a central part of our lives, and because it is bound up with such deep and intimate experiences. Comparisons with prostitution are made when surrogacy is commercialized.

Intimate and commercial relationships do not fit together easily. In relationships between friends or lovers, or in families (at least as they should be), we confront each other without calculating commercial gain, and do not assess each other mainly in terms of wealth. In families and between friends, gifts are more common than sales. This antagonism between intimacy and commerce is part of the reason why prostitution does not seem the ideal model of sexual partnership, and why other forms of commercialization of sex are tolerated rather than admired.

No doubt it is impossible to prevent individuals paying others for sex; but, if a large company set up as an agency for prostitution, this might seem an unacceptable further step towards turning sex into a commodity. In a similar way, commercial surrogacy agencies can be seen as contributing to a society in which parenthood is seen as another commodity. And, because relationships are partly constituted by how they are seen, this threatens an unwelcome change in the relationship itself.

Because the commercialization of intimate relationships seems something to resist, and because restricting the liberty of commercial organizations seems less intrusive than restricting the liberty of individuals, we do not favour permitting commercial agencies for surrogacy.

The question of commercial agencies may be one of the few

we have considered where there is room for a distinctively Western European approach. In the United States, there is a strong presumption in favour of the free market, while the countries of Eastern Europe operate with a strong presumption against it. Generalizations about people living in large geographical regions are obviously suspect, but in Western Europe there does seem to be a strong current of opinion favouring a society between these extremes. We think that the policy of discouraging payment to surrogates without imposing a legal ban, and of banning commercial agencies, fits this approach.

Part Four
The Unborn and Research

Chapter 8
The Unborn

Some of the possible benefits of the new technologies depend on using means which accentuate disagreements — familiar from the abortion debate — over the status of the 'unborn child'. The approach we develop to the status of the 'unborn child' has applications to issues discussed later, such as the question of terminating pregnancies where tests pick up serious abnormalities. But in this part of the report we focus on the issue of research and transplants.

The techniques of *in vitro* fertilization require, for their most effective use, the creation of more embryos than are intended to develop into babies. The first question is whether we are entitled to create them in this way. Once we have done so, there are questions about whether we are entitled to dispose of them in a way that kills them before they develop, and also whether we are entitled to use them for research purposes. If we say 'yes' to any of these questions, there is the problem of how far along the developmental continuum we are willing to intervene.

Some disorders, such as Parkinson's disease, may be treatable using material from aborted foetuses. Is this acceptable? What is the morality of taking material from living embryos or foetuses in order to grow it up for transplanting?

1. A note on terminology

Discussion of such issues as 'embryo research' is not helped by widespread vagueness about the stage of development to which the term 'embryo' applies. We proposed to follow current medical and scientific usage, until we found large discrepancies of usage in the scientific and medical literature. With some caution, we intend to use three key terms as follows:

Pre-embryo: the product of conception, from fertilization to nidation, i.e. up to implantation in the uterine cavity. This stage lasts about 10 days.

Embryo: the product of conception during the stage from implantation in the uterine cavity up to six weeks after fertilization.

Foetus: the product of conception from the end of the embryonic stage to birth.

So far in this report we have talked of 'embryo' donation and transfer, rather than of 'pre-embryo' donation, etc. Because the techniques in question are only relevant at particular developmental stages, we have not up to now needed the additional precision given by this distinction.

It should be stressed that none of these definitions is intended to prejudge any moral issues. For instance, any right a pre-embryo may have to life is not diminished by calling it a 'pre-embryo' rather than an 'embryo'. And our occasional use of the term 'the unborn child' to include all stages of development up to birth is not meant to *entail* the controversial claim that a foetus has the rights of a child. 'Unborn child' could be a term like 'unconceived child', with no implication that a child already exists. Moral issues are not determined by names. And our intention is not to adopt terminology slanting the discussion in either direction.

2. The question of rights

Opponents of therapeutic abortion or of embryo research need not deny that there are benefits to be derived from them. They say that the benefits do not justify overriding the claims of the embryo or foetus.

There are different possible claims on behalf of the unborn child. If there is a right to life as far back as conception, this is violated both by abortion and by the disposal of embryos after they have been used for research. There is also the objection that abortion may cause pain. And it is said that the embryo has the right of being treated with a respect which rules out using it for research. But the biggest issue is the right to life.

It is widely held that unborn children, like other humans, have a right to life. Claims about the right to life, or indeed about rights in general, are often vague or ambiguous, and need clarification. They are not normally about legal rights: those who campaign against laws permitting abortion are often saying that the foetus has a moral right, which is not, but should be, recognized by the law.

We have seen that such claims about moral rights need both clarification and justification. Are such rights thought of as absolute: inviolable under any circumstances? Or are they something weaker: providing a presumption against killing the foetus, which can perhaps be overridden in exceptional circumstances, such as pregnancy after rape, or a threat to the mother's life? How do we decide which human interests are so basic as to generate *rights*, rather than lesser moral claims? But, whatever the answers to these questions, it is likely that the preservation of life would figure in most lists of basic human rights.

In this chapter, we will use the terminology of rights, to avoid repetition of more cumbersome phrases. But this should be understood as being subject to the reservations already expressed.

The arguments in favour of the view that the embryo or foetus has a right to life fall into two broad categories:

Appeals to the present status of the unborn child.

Appeals to what it has the potential to become.

3. The present status of the pre-embryo, embryo or foetus

Two claims are not always distinguished. It is said that a pre-embryo, embryo or foetus is a person, and alternatively that it is a human being.

The view that it is a person raises the question of the criteria used to decide what counts as a person. There is no consensus about this. Attempts by philosophers to draw the boundaries of the concept of a person have been notable for their divergence. The boundary has sometimes been drawn very broadly in one direction to include the newly fertilized egg, and very broadly in other directions to include members of other animal species, Martians and robots. It has been drawn more narrowly to include only those who are self-conscious. Daniel Dennett has suggested that having a sense of justice may be a necessary condition of being a person. (*Brainstorms*, Hassocks, 1979, p. 282.) This philosophical diversity may reflect a lack of agreement among all of us who make use of the concept.

Our everyday concept may be indeterminate in a way that makes it unhelpful in the abortion issue. Being a person may not have sharp boundaries. On this view, just as there is a blurred boundary rather than a single moment when we become middle-aged, so we only gradually acquire the distinctive characteristics of personhood. There is a tendency for supporters and opponents of abortion to stipulate a definition of 'person' that supports their own views. But it may be too optimistic to expect that any such boundary-drawing will correspond to any already-agreed and sharp conceptual frontier.

The issue of whether the embryo or foetus is a human being may seem more promising. No-one denies that it is alive, and it is surely a member of our species rather than of any other. But the problem with this argument is that it applies equally to the unfertilized egg or to the human sperm cell. This argument easily enough proves that the embryo or foetus is a human being, but it is not clear that the status of 'human being' in this minimal sense brings with it any moral rights. It is widely assumed that qualifying as a human being is sufficient to guarantee the possession of a right to life. But this assumption is questionable, and perhaps derives much of its plausibility from our thinking of 'human beings' in terms of our friends and neighbours. An embryo is not the kind of human being you can share a joke with or have as a friend. There is a problem in saying what it is about the *present* state of an embryo that gives it any stronger moral claim on us than, say, an oyster.

4. Potential

Perhaps what matters is not some property an embryo now has, but what it has the potential to become. This claim seems to imply that disposing of an embryo is wrong because it prevents the existence of a particular developed person, namely the one the embryo would have become. But this argument rules out contraception. You are a particular developed person, and contraception would have prevented your existence. The apparently innocuous word 'potential' turns out to be very slippery.

To avoid ruling out contraception, potentiality has to be interpreted differently. Perhaps the destruction of potential is not just a matter of the loss of a future developed person, but also the loss of something that has got a certain distance on the way there: the programme is already in existence, as all the genes are present. But this leaves some unanswered questions. If the properties an embryo now has do not generate a right to life, and the argument about one less future person does not do so either, why should combining the two considerations give the desired result? Of course a compound can have properties not possessed by its individual ingredients, but in this case some account is needed of the moral chemistry involved.

The conventional assumption, that stopping the process of creating a child is morally different according to whether it is done before or after conception, may be linked to other beliefs we have about the importance of a person's identifiability. We care more about rescuing a single trapped miner than about improving safety standards in mines in ways that would save larger numbers of miners in future; society often spends more money on a single dramatic rescue than on measures that would save many more 'statistical' lives.

This discrepancy may involve a kind of irrationality. (Suppose the dramatically rescued miner next month becomes one of the ten 'statistical' lives claimed by the poor safety measures?) But, whether this attitude is rational or not, it is rooted in a natural reluctance to turn down an appeal to save the life of an identifiable person who is at this moment desperate to be saved. It may be that this greater concern for an identifiable person carries over to reproductive issues, making even the earliest abortion seem qualitatively different from contraception. But it is worth noting how different the reproductive cases are. In the absence of consciousness, there is no-one desperately hoping to live and no family who will be devastated by the loss of someone they care for. When we carry over our emotions towards identifiable individuals to the unborn, we are leaving behind the context from which those emotions perhaps derive their justification.

Part of what makes parents want an embryo or foetus to be born is the projection they make of his or her person into the future. That may explain why parents, when a miscarriage or even an abortion takes place, sometimes feel they have lost a child. But from this, it does not follow that the embryo or foetus *is* a child.

5. Some suggested boundaries

Because development in the womb is more like a smooth curve than like a series of quantum jumps, those who support any embryo research or abortion have the problem of drawing a moral boundary between what is and is not acceptable. It is often argued that, unless we draw the line at conception, there is no stable and defensible boundary until birth. There are grounds for doubt about all the sharp boundaries which have been defended by different parties in the abortion debate.

(a) Conception: At conception, there is for the first time a single body, which will have a continuous history through the pregnancy and beyond: through an entire lifetime. The moral weight given to this will depend, among other things, on how far bodily continuity is taken to be a *sufficient* condition (rather than merely a necessary condition — itself a debatable claim) for the identity of a person or human being.

The assumption that conception is the boundary may be the basis for thinking that IVF solely for the purpose of research should be prohibited. (This view was taken in West Germany both by the Benda Commission and by the Medical Association. But they suggest that, where there already exist pre-embryos which are not to be implanted, there should be a fourteen-day limit to research. Taking the right to life more strictly would have required that such pre-embryos should be frozen and implanted on another occasion.)

Drawing the boundary at conception may be criticized. It is not clear that it is the sharp boundary people have hoped for. The 'moment' of conception may be an illusion, as the position is complicated by the 'two pro-nuclei stage', which continues for a number of hours. This is the stage just after the ovum has been penetrated by the sperm cell, but before fusion of the two nuclei. There are still two nuclei with twenty-three chromosomes each, rather than a unit with forty-six chromosomes. Biologists working with animals have been able to change the genetic composition of the fertilized egg by intervening at this stage: by extracting one of the two nuclei and replacing it with another. So perhaps there is no clearly defined human or animal entity at this stage.

These difficulties need not be decisive. It could be argued that the right to life begins either when the ovum has been penetrated or when fusion is complete. But neither of these as a moral boundary has the apparent obviousness of what was once thought to be the single moment when the sperm cell and the ovum met and merged.

There is also the objection that the plausible reasons for holding that people have a right to life, such as consciousness, or the desire to live, are not remotely present in the pre-embryo.

(b) Nidation: Another boundary which has been considered is nidation, or implantation, taking place at about ten days. But again, it is hard to see that the embryo has a stronger moral claim than the pre-embryo. Why should the fact of being implanted in the uterus be thought so important?

(c) The formation of the primitive streak: It has been held that the formation of the 'primitive streak', at about fifteen days, is the point at which rights begin. (The Warnock Committee suggested that, in the United Kingdom, research on embryos should be forbidden after fourteen days.) One argument for this is that, up to this stage, more than one individual can be formed, while no such indeterminacy remains later. But this boundary too can be questioned. Why is the number of individuals, or potential individuals, relevant to what rights it or they have? If division could occur much later, say at five months, would the onset of rights be deferred until then? If so, why? It is unclear what the reasons are which prevent the choice of potential divisibility from being an arbitrary one.

Another argument for taking the formation of the primitive streak as the boundary is that it is linked to the beginning of the nervous system, and so to the possibility of consciousness. The formation of the primitive streak is at best the precursor of the developed nervous system which is the basis of consciousness. Part of the problem is that we have little understanding of what degree of neural complexity is required for consciousness. But the underlying issue is whether the onset of consciousness, whenever that is, should be taken to be the boundary.

(d) Consciousness: This proposed criterion is the point where enough of the nervous system has developed to make it likely that consciousness is present. This seems highly relevant to a right not to be caused pain, but it is not clear that it is sufficient to generate a right to life. Many animals killed for food, or used for experimental purposes, give every sign of being conscious. Perhaps they too should be accorded the right to life, but until we do this, it is hard to maintain that mere consciousness, however primitive, is sufficient to give humans this right.

(e) Viability: The point where the baby could survive outside the womb is sometimes taken to be where the right to life begins. One problem is that this is a shifting boundary. It makes foetal rights depend on the current state of medical technology, and so may not be the deeply significant change we are looking for. There is something odd about a frontier according to which

two foetuses at the same stage may vary in whether or not they have a right to life according to how their local hospital is equipped.

(f) Birth: Placing the boundary at birth can be challenged. We make great efforts to keep premature babies alive. Why should a baby at the same stage of development have no claim on us just because it is still in the womb?

6. The ascending slope view

There is an alternative view, according to which there is no sharp boundary, but a continuum, such that acts quite acceptable at the start of development become increasingly hard to justify as later stages are reached. On this view, the problem is differently conceived. For legal purposes, the boundary may have to be sharp, but this will not be seen as reflecting any sharp change in development. The model will be something like the speed limit, and what will be important will be to draw it in roughly the right region, reflecting the various factors we think important, rather than hoping we have correctly guessed the moment when being human or being a person 'really' starts.

It is worth noting that this approach does not necessarily draw all the moral frontiers in one place. Concern to set limits to embryo research is sometimes dismissed on the grounds that, if abortion is allowed, then concern for the embryo is misplaced. But if there is no metaphysical frontier, there is the possibility that abortion and embryo research should be treated differently, perhaps because the reasons supporting them are of different weight.

Chapter 9
Research and Transplants: The Possible Benefits

What is the case for research on pre-embryos and embryos? It could bring great benefits to the treatment of infertility and other reproductive problems. It could also help in the understanding and treatment of other disorders. For instance, it gives a chance of understanding diseases where cells grow too quickly, such as cancer. We may come to understand the origins of cancer by looking at genetic mechanisms in embryonic cells, and seeing how the genes that control an abnormal cycle-time express themselves. Another possibility is that cells taken from embryos and foetuses will help in treating such conditions as Parkinson's disease.

1. Gene defects

Genetic disorders are remarkably common and can be disastrous for those who have them. Embryo research is now opening up the possibility of developing techniques for screening for a wider range of these disorders. And, in the longer term, there may be the possibility of genetic intervention to treat them. We leave discussion of the issues these developments raise until the final part of our report. But, in thinking of what benefits research may bring, the extent to which gene defects can now ruin people's lives should not be forgotten.

2. Infertility

We have noted how devastating an experience infertility can be in a couple's life, and how frequently couples run into difficulty trying to conceive. With all our present techniques, only about a third of those treated for infertility manage to conceive. IVF only helps a very few. We need to know more about the causes of infertility. And, related to this, we know that only between eight per cent and 16 per cent of the embryos we put into the uterus are implanted. To improve on this, we need to study the interaction of the embryo with its environment. It is also desirable to develop techniques for assessing how well an embryo is developing.

(a) Implantation: Before implantation, the embryo sends chemical messages to the uterine lining and receives messages back, enabling it to choose a site for nidation. To understand this process and the ways it can go wrong, it would be helpful to study it *in vitro*, using an embryo and a section of uterine lining.

(b) Sperm: Twenty to thirty per cent of infertility is caused by problems with sperm. About ninety per cent of male infertility is virtually untreatable. Most infertile men produce defective semen, containing abnormal sperm cells. With men who produce abnormal sperm, we can only hope that a fair number of normal sperm are also produced. We need research on the genesis of sperm. But we also need to identify those normal sperm that *are* produced in these infertile men, make sure that they can come into contact with an egg, and then check that the fertilized egg is normal. Research on this requires the creation of embryos, which it would be unwise to return to the uterus.

3. Contraception

Perhaps one obstacle to dealing with the population explosion is the risk of side-effects carried by current methods of contraception. Attitudes towards using them may change as the risks diminish. The next generation of contraceptives will operate by preventing fertilization, preventing the development of the early embryo, or preventing implantation. Embryo research is important for the safety of these methods. Because they will fail in some cases, it is necessary to find out whether they then have any tendency to produce abnormal embryos or later ill-effects.

4. Miscarriage

Miscarriage, especially when repeated, can be a severe blow to a woman, who may view it as the loss of a child, and it is fairly common. (There are about 100,000 hospital admissions a year for it in the United Kingdom.) The main causes are defects in implantation or in early embryonic development. It is desirable to look at embryos growing *in vitro* to study the normal line of development.

Twenty-five per cent of cultured embryos develop abnormalities. This may well also be true of embryos growing naturally, with miscarriage a way of filtering them. One way of approaching the problem of miscarriage would be to study the embryos of those prone to it.

5. Ectopic pregnancy

Sometimes an embryo implants in an abnormal site, such as a tube or an ovary. These ectopic pregnancies are quite common. In Europe they occur about once in every 150 pregnancies. Elsewhere they are more common: in Jamaica they occur once in every 20 pregnancies, and they are even more common in parts of Africa. They cause some deaths in Europe, and in some parts of the world they are the commonest cause of maternal death. Embryo research holds out the possibility of better understanding of the implantation mechanisms, which are one possible cause of ectopic pregnancy.

6. Egg banking

A number of cancers in young people can now be treated by chemotherapy and radiotherapy. An instance is leukaemia in young women. But the treatment makes them sterile, by killing germ cells in the ovary. With the availability of egg-freezing, women will increasingly ask for their eggs to be banked before the treatment, so that they can still have children. If this is done, we need techniques to check that the embryos are normal.

7. Transplants

There are possible benefits from growing cell lines for longer in order to study their differentiation into nerve, muscle and other tissue. This might also enable us to use the products of such cell lines to replace damaged tissue in heart, kidney or nerves. We might be able to use embryonic blood products from a cell line: bone-marrow cells might be grown for people with leukaemia, perhaps bypassing the problem of rejection.

Monkeys with drug-induced Parkinson's disease have been 'cured' by transplants of brain cells from monkey foetuses. (This work was described by Professor D. Eugene Redmond, at the Sixth International Catecholamine Symposium, held in Jerusalem in 1987, whose *Proceedings*, edited by M. Sandler *et al.*, are in press.) The same thing in humans, using tissue from aborted foetuses, has been attempted in Sweden, Mexico and Britain. Only cells still at the stage of dividing will transplant and grow. So foetal cells, rather than those from recently dead adults or children, are needed. The evidence suggests that the transplant will only work with cells from foetuses of not more than nine weeks.

The procedure is at a highly experimental stage of development. The extent of its success is unclear. But, for the future, it may be one of the most promising approaches to very severe Parkinson's disease. And it holds out similar hope of treating, for instance, a disrupted spinal cord.

8. The case for research

If these various kinds of research are not vetoed by some other overriding ethical consideration, this field is exceptionally rich in possibilities for good. When embryos created in the course of IVF are not implanted, these possible medical benefits suggest to some doctors and scientists that it would be unethical to dispose of them *without* using them for research.

Chapter 10
The Ethics of Research and Transplants

The case for allowing some research on pre-embryos and embryos is a strong one. As always with research, the results cannot be guaranteed: they may be either less or more impressive than first hoped for. But the possible benefits are substantial. However, they cannot be considered in isolation from the means by which they are obtained. On one view, these means are comparatively unproblematic: the research is similar to that on sperm cells. On another view, this research is on a child, and should be as restricted as research on, say, a five-year-old. The status of the pre-embryo or embryo is crucial. How should the ethical issues be approached?

We start with two restrictions which we hope will be relatively uncontroversial, before turning to more divisive issues.

1. The exclusion of research causing pain

When a foetus has reached the stage where it can feel pain, research causing pain should not be carried out. Whether or not the foetus has other rights (to life, etc.), the only condition relevant here is that it *can* feel pain.

There are problems about the kinds of facts about behaviour or neural complexity which are to count as evidence for foetal experiences. But potentially painful research should be excluded at whatever stage claims about pain start to look plausible. It is desirable to err on the side of caution.

This exclusion does not apply to research on the pre-embryo. The primitive streak, from which the nervous system develops, is formed only at about fifteen days. And there seems very little reason for thinking that even the late embryo, nearing six weeks, has the neural complexity to support consciousness. But, erring on the side of caution, we may begin to be concerned about pain around the start of the foetal stage.

2. The exclusion of research harming the child who will be born

No-one thinks harmful research on new-born babies is acceptable. Research doing the same harm to the new-born baby does not become acceptable just because it takes place earlier. What matters is the harm to the later child, not the time it is caused. Any such research on pre-embryos or embryos should be excluded.

Thus we must exclude interference with aspects of normal development which are needed for later physical or psychological well-being. Take rearing a child entirely outside the natural womb. In normal pregnancy, chemical and other links between mother and baby may be important in ways we do not yet understand, and which we would probably not succeed in reproducing in an artificial womb. This suspicion is a strong objection to the procedure.

Notably, objections based on the risk of harm to the later child do not apply to research where, because the embryo will be destroyed, no baby will be born harmed. This raises two much more controversial issues: the right to life, and the respect due to the pre-embryo or embryo.

3. The right to life

There are three possible views about the right to life of an embryo at a particular stage. One is that it has the same right to life as a developed person. The second is that it has no right to life. The third, intermediate, view is that its life has a moral claim on us, but a weaker one than it will have later.

We have already looked at some arguments for the 'full strength' right-to-life case. Most of them are intended to draw the moral boundary at 'conception', or some more precise point within the scope of that rather blurred term. An embryo is in a certain sense a 'human being', but deriving a right to life from this is problematic. The morally relevant features which support a right to life in a developed person (consciousness, the desire not to die, etc.) are not present in the embryo. An alternative derivation of the right to life is from what the embryo has the potential to become. But we have seen that this line of thought slides towards the unwelcome conclusion that contraception and indeed chastity are wrong.

It is implausible that embryos have a full strength right to life. The convincing and unproblematic arguments which would be needed to support the claim have not yet appeared.

Some members of the committee take the 'ascending slope' view. According to this, the life of the pre-embryo has some claim on us, a slight claim to start with, but increasing during the developmental process. Others think that, even at the end of the embryonic stage, few or none of the requirements for having a right to life are satisfied. The ascending slope view may be right, but the line has not yet started to slope upwards.

In practice, the two views come close to convergence. They both suggest that research should not be vetoed automatically when it leads to the destruction of a pre-embryo or embryo. On one view, this creates no presumption against the research. On the other view, it creates a weak presumption against the research, which has to be shown to be sufficiently beneficial to justify the destruction of embryos.

4. The right to respect

Some people's view of this research is based on respecting the pre-embryo and embryo. The idea of respect for persons, or for human beings, has strong roots in the European ethical tradition. It is found in Judaism and Christianity, and has its most powerful secular expression in the philosophy of Kant. It is the principle that we should treat people not merely as means, but as ends in themselves. We should respect their dignity, rather than treating them in ways that degrade them.

This view clearly captures something important in our attitude to people. But there are problems in seeing exactly what is ruled out by such a principle.

On one interpretation, people have certain interests which are so important that they should be respected as rights. But, as we have seen, this raises questions about what rights we have, and about how absolute they are.

On a second interpretation, respect for persons involves respecting their autonomy. It involves, for instance, not carrying out medical experiments on people unless they have given free and informed consent. This might be thought to rule out embryo research, as embryos cannot give consent. But it could be said that respect for autonomy is only owed to people, or to those capable of taking decisions, and that on neither ground do embryos come within the scope of the principle.

On a third interpretation, respect for persons is a matter of according them a certain dignity. This includes showing consideration for them in relationships, and it can extend beyond the treatment of people who are capable of consciousness. For instance, the mutilation of a corpse might show lack of respect. The reaction against the idea of people setting up as dealers in embryos may be related to concern about this kind of respect. So may be the revulsion many feel as the thought of foetal material being used in industrial processes. And reports of doctors using neural material from aborted foetuses may trigger a similar response in some.

5. Psychological barriers

Perhaps the most important aspects of actions are their effects on people and their experiences. Some of these psychological reactions to lack of respect may seem irrational. Corpses and embryos are not aware of anything, so does lack of respect to them really matter?

The role of these psychological reactions in ethics is disputed. Some people think that they are the basis on which we construct our morality, that our feelings of approval or revulsion are what conscience consists in. These 'moral intuitions', as philosophers rather grandly call them, are viewed as the basis of ethics. On another view, these moral intuitions are merely 'gut reactions', and rationality in ethics consists in standing back from them and asking whether or not they are based on defensible ethical principles.

Each of these views has some truth. It is hard to imagine that humans could have much interest in an ethical system quite divorced from human responses. But reflection on some emotional reactions can lead to them being justifiably rejected as a basis for policy. Learning to become a surgeon, for instance, involves overcoming a natural revulsion against cutting open the human body.

In the case of foetal research, some psychological barriers should be protected. Few changes in attitude could be more appalling than to start thinking of newborn babies as disposable material for research. The later foetus is not very different from a baby, and it might be very hard to retain the normal response to babies while breaking down the barriers to late foetal research. This aspect of *us* is a good reason for being much more reluctant to support later research than embryo research, independent of any question of foetal rights at different stages.

Perhaps this reason also plays a part in our concern to rule out experiments involving foetal pain, which need not be worse for the foetus than what is involved for animals in similar experiments.

6. Transplants

Perhaps neural material from aborted foetuses will be able to reverse very severe Parkinson's disease. The potential benefits are obvious. But what of the ethics?

(a) The right to life: Do such transplants violate a foetal right to life? The defence against this charge usually given is that the foetuses are aborted, and so their lives have not been lost because of the transplants. Those who think that abortions violate a foetal right to life will not think that such use of foetal

cells mitigates the violation. But it is hard to see how those who accept that abortion can be justified could object, on grounds of the right to life, to the use of these cells from already aborted foetuses.

There is the question of whether the foetus is dead at the point when the transplant is taken. The brain cells must be alive if they are to be any use. A 'whole brain death' criterion of death is usually accepted in other transplant cases: the death of the brain stem is used as the test. Should the presence of living brain cells count against the view that the foetus is dead?

There is a distinction between an organ being alive and there being life in some of the cells which make it up. Some people favour an 'upper brain' definition of death. The crucial feature of this definition is that it excludes any recovery of consciousness. Conscious life is over, and this is compatible with particular brain cells still being alive. This distinction can be applied to the foetus, but the point here is that conscious life has not yet started. These foetuses, between six and nine weeks old, are not at the stage where it is at all plausible to think conscious life has begun. Many who would not go so far as accepting the 'upper brain' definition of death would still accept removing medical support from someone who had irreversibly lost consciousness. The claim is that this would not violate the right to life, which is bound up with possession of consciousness. A parallel point can be made in the case of the foetuses to be aborted at this stage.

(b) The right to respect: Those who accept that the abortion was justified may still feel that to use the brain cells for transplants denies the foetus proper respect. But this view is difficult to reconcile with accepting transplants in other cases. If respect for the corpse of a dead adult is compatible with taking organs for transplant, why should things be different in the case of a foetal corpse?

(c) Psychological reactions: The use of foetal material arouses revulsion in some people. Is this a strong objection to these transplants?

The emotional reaction has some likely sources. One is the

protective response aroused in us by babies. Another is the feeling that bodies should be respected.

Undermining the protective feeling towards babies would be a disaster. The feeling might be threatened by painful or mutilating experiments on living foetuses. But, although the feeling is aroused in some by the transplants, it seems unlikely that this use of non-viable foetuses would change our responses to living babies. Organs are removed from people with donor cards who die. It is unlikely that the surgeons involved come to care less about ordinary people. Taking organs may also go against intuitive feelings of respect for bodies, but we count people's need for replacement kidneys or hearts as a lot more important. The lives of living people are wrecked by Parkinson's disease. With foetal transplants, it seems perverse to place respect for bodies before rescue for the living.

(d) Pressure on women to abort: There seems a real danger that a pregnant woman, perhaps uncertain whether to continue with the pregnancy, might be subject to pressure to have an abortion. It is hardly likely that such pressure would arise merely in the abstract: "Have an abortion: someone would benefit from the brain cells". The pressure can more realistically be thought of as coming from a link with a particular Parkinsonian patient. The need to avoid this is a good reason for a rule excluding transplants where there is a link between the woman carrying the child and the beneficiary.

(e) Creating foetuses for transplants: There is another possible source of the reaction against the transplants. We may be anxious about what horrors could lie further down this road. There are thoughts about people conceiving children in order to use them as organ banks at the foetal stage before killing them.

The Mexican scientists doing foetal transplants for Parkinson's disease deliberately used a foetus which had spontaneously aborted. One reason for this could have been opposition to induced abortion. Another possible reason would have been to keep some distance between inducing abortion and deriving benefit from it. This would fit with anxieties about sliding into a policy of conceiving children solely to be organ donors.

These anxieties also arise more generally about embryo research. The benefits of research will often reasonably be thought stronger than a case against it based on claims about the embryo's right to life or to respect. But there may be disagreement on the further question of whether it is acceptable to create embryos specifically for purposes of research.

Some think this flouts the Kantian principle of treating human beings always as ends in themselves and never merely as means. Others think that, if it is not wrong to use an already existing embryo for research and then to destroy it, it is hard to see why it is worse to create an embryo in order to use it in the same way. A similar sceptical view could be taken about the importance of the distinction between using existing foetuses as donors and creating them for the same purpose.

Attitudes on this question will vary according to views about the right to life, about the kind of respect due to embryos and foetuses, and about the content of the principle forbidding treating human beings merely as means.

The creation of embryos for research presupposes people willing to donate semen and eggs for research purposes. Some semen donors do donate for research, in which case they are always paid. But can the same thing be expected of egg donors? If not, some women, perhaps IVF patients, would have to be superovulated to create some extra eggs for research. This raises questions about the interests of such women, and about the necessity for obtaining their consent, which need to be answered before such policies are embarked on.

There is also a question about what we would be doing to ourselves in arranging conception with no hope of a child, but simply to use and destroy the embryo or foetus. This seems a large step towards viewing embryos as commodities, possibly bringing with it an erosion of our respect for later human beings.

Perhaps we could adjust to this policy, as we have to abortion, without becoming dehumanized, and without damage to the subtle psychology bound up with the creation of our children. But there is at least a question here, which gives a reason for caution to those with no belief at all in a foetal right to respect. In this regard, there is something less chilling about IVF in the laboratory for these purposes than about a woman getting pregnant in the normal way for the sole purpose of giving her husband a transplant. If what matters is our psychology, this second case is more objectionable. We think that transplants should be acquired from foetuses aborted on other grounds, and that the beneficiary should normally be someone unknown.

The extraction and growth of a line of embryonic cells of a particular type, without the development of other parts of the body, could provide the basis for replacing damaged tissue in the heart, or for replacing bone marrow cells in leukaemia (possibly without the rejection problem). This would be the culture of cells and organs, rather than human beings, and would not be open to the ethical objections which apply to later foetal research. Of course, those who find any research on pre-embryos unacceptable will find human cell culture objectionable too.

7. Time limits

In setting limits, the important aims should be to make sure that research stops before there is any likelihood of pain being experienced, and before the foetus starts to resemble a baby enough to arouse the same emotional responses. It is apparent that neither of these considerations easily dictates a sharp boundary. Any boundary drawn on this basis will have the partly conventional aspect of a speed limit. No-one thinks there is some magic about sixty kilometres per hour which makes it different in kind from driving at sixty-one. The important thing is to set the limit in broadly the right place, recognizing that the precise point has a degree of arbitrariness. There is a case for a body which has the power to set time limits to research, and which would, like the traffic authorities, be able to revise its rules in the light of new evidence.

Chapter 11
Monitoring

The crude 'technological imperative', which states that if something can be done it will or even *should* be done, is not an intelligent approach to the application of science. The problems of the first industrial revolution or of the development of nuclear weapons were left to be dealt with afterwards. Fortunately, the wrongness of this approach has become clear before the revolution in human biological technology is fully under way.

But there are unsolved problems about harnessing this revolution to human values. One approach would be to impose external regulation on scientists and doctors. Against this, there is a case for their autonomy. They may fear research being stifled by scientifically ignorant bureaucrats. And there are questions about how effective external control is likely to be. But, on the other hand, the impact of work in these fields is likely to be so great that a policy of 'leave it to us' may strike non-scientists as dangerously complacent.

There are two related aspects of these issues. One is that, if non-specialist members of the public are to have a voice in what is done, the level of public information and debate needs to be raised. The other is the need for mechanisms by which research and its applications can be guided by the values of the wider society. This applies to embryo research and to the development and use of such techniques as gene therapy.

1. Information and public debate

Debate about medical or research ethics is shaped by the chain of scientific information. Information hardly ever goes directly from the scientist to the public. Scientific results in journals are taken up — highly selectively, on grounds of news value — by the media. A few striking items can create an impact out of all proportion to their importance, creating either false hopes of cures or demands for the banning of disturbing new techniques. Public opinion may swing between wanting stringent control and the escapism of leaving things to the experts.

Rational and informed public debate encourages a more discriminating evaluation of research and its applications. There is no magic formula for this. It will emerge only out of slow, incremental changes of attitude. Perhaps it *is* emerging out of them. Scientists are perhaps more willing to engage uncondescendingly in public debate and to warn about likely problems. We are beginning to receive a flow of intelligible and reliable information about results and their importance; and perhaps the scientific habit of suspending judgement before obtaining evidence is starting to spread.

It is easy to be in favour of these attitudes conducive to a better public debate. But they will only develop with practice: with public discussion and media attention. They will also develop with the increasing use of public hearings. (In this, the United States is far ahead of most European countries, though we may want to adapt their example to our own less legalistic and less adversarial traditions.) On major matters, an official committee (preferably with something like equal numbers of specialist and lay members), followed by a Parliamentary debate with a view to possible legislation where appropriate, can transform public awareness.

It is not always enough to have public debate and then leave all the decisions to individual scientists or doctors. But any regulation should be based on something less heavy-handed than the law.

2. Ethics committees

At the level of individual research projects, ethics committees are such an instrument. These arose partly out of new medical techniques, and the need to protect human subjects in research. Their development was influenced by the Nuremberg Code and by the Declaration of Helsinki, which required that 'all experimental protocols should be transmitted to a specially appointed independent committee for consideration, comment and guidance'.

Ethics committees protect the rights and interests of subjects of research. They also look at the scientific merit of a study. Scientific standards are obviously crucial. A poorly designed study uses resources, and perhaps involves risks, while giving no knowledge in return. The ethics committee's view of the ratio of risk to benefit should be more impartial than that of the researchers.

Ethics committees also consider moral questions related to the practice of medicine. For instance, a doctor wondering whether to help an unmarried woman or a lesbian to get access to AID may find the deliberations of ethics committees about similar cases helpful. Perhaps there will be no final agreement on such matters, but the consideration by different committees of a range of cases may help us towards some degree of consensus.

How well do ethics committees filter out what is unacceptable? Their authority is rather ill-defined. They have normally been set up on the recommendation of medical or scientific bodies, or of Ministries of Health. In the United States, federally funded research has to be scrutinized by Institutional Review Boards. In other countries, the views of the ethics committees are often backed by the witholding of funds or research facilities. And a number of medical and scientific journals only publish results of projects approved by ethics committees. There is also moral pressure: many doctors and scientists would be reluctant to go ahead with a project turned down by an ethics committee. The authority of such committees is greater where they are not in effect self-appointed, but have an official status and democratically appointed lay members. For questions about having children, though not only for them, there is an overwhelming case for women having a substantial presence.

Despite the slightly precarious authority of these committees, they are our most promising mechanism for bringing ethical appraisal to bear on research. They are less rigid than the law, and yet have some sanctions. They carry more moral weight than purely informal expressions of individual opinion. This moral authority is enhanced to the extent that reasons for past decisions are a matter of public record, and the equivalent of case law is built up. Decisions of different committees (or of the same committee at different times) may not always be consistent. But out of many decisions a tradition will emerge, and this tradition may be a plural society's best approximation to consensus.

3. Licensing authorities and other regulatory bodies

Ethics committees are a tactical rather than a strategic approach to guiding research. Some of these issues are of such importance, and arouse such strong feelings, that central social decisions are required. The stage at which embryos or foetuses have claims which override those of research, for instance, seems too large an issue to leave to a multitude of local decisions. Some who take this view favour laws either banning all embryo research or else setting some defined limit to it.

But there are middle courses, which avoid the rigidity of law, while still having a general social policy. One would be a series of *ad hoc* commissions, created as new ethical problems appeared. But this would be time-wasting and inadequate where there is likely to be a steady flow of ethical problems. A better alternative is a permanent regulatory body, to develop and publicize criteria for acceptable experimentation, and to look at proposed projects. This body might be given the job of licensing all experiments in the field. Or it might lay down guidelines for ethical committees, with the understanding that the committees should refer directly to the regulatory body any proposals which raised new issues of principle.

In the version favoured by the Warnock Committee, the regulatory body is a statutory one, with legal powers to control research in these fields. In Denmark there are seven regional research ethics committees, which together with the Central Research Ethics Committee, have legal recognition. In 1987, this system was extended by an act setting up an Ethical Council, containing scientists and lay people, set up mainly to deal with those problems of research and techniques which relate to human procreation.

To have adequate authority, the regulatory body should be statutory, but should not necessarily have the legal power to license or forbid projects. The alternative of laying down guidelines for ethics committees may have advantages.

The French experience is perhaps relevant here. In 1983 a decree set up the Comité Consultatif National d'Éthique pour les Sciences de la Vie et de la Santé. The President of France appoints five members: a president and four members representing different religious and philosophical outlooks. (The idea of philosophical outlooks being represented on a par with religions perhaps reflects a culturally distinct conception of philosophy. But the system could easily be adapted to cultures in which the role of philosophers is more analytical than systematic.) Fourteen other members are chosen, according to specific statutes in the decree, for 'their competence and interest in ethical problems'. (Some of these are members of the medical profession or scientists.) A final fourteen members are chosen for their competence in research. The president has a renewable two-year appointment. Half the other members are replaced every two years.

The French National Committee, although statutory, has no legal powers to implement its decisions. It can propose legislation, but the proposals can be rejected. In 1985, the French government decided not to follow a committee recommendation for a new law regulating experimentation on human subjects; and in 1986 the government decided not to follow the committee's recommendation for a three-year moratorium on certain aspects of embryo research. The members of the committee are divided on whether this merely advisory status is satisfactory. Some see it as undermining its authority. But the committee's president, M. Jean Bernard, has said that its advisory status is what guarantees its independence.

The powers desirable in such a committee will vary according to the relative weight given to independence as against efficient and rapid intervention in issues as they come up.

The regulatory body should be independent of government, and should contain about equal numbers of scientific and non-scientific members. The aim would be for it to be both informed and independent. Women should again be well represented.

The regulatory body should have the power to commission relevant social research. The social effects of scientific developments are crucial to the ethical issues, and decisions are often based on too little information. If the body regulated not only research but also other aspects of assisted reproduction, research on the effects of being brought up by a single parent or by a lesbian couple, or about the effects on children of donor anonymity would obviously be highly relevant. Members of the body will presumably change their minds in the light of scientific or sociological evidence, and will see the need for obtaining it.

The regulatory body should publish reasons for its decisions, just as ethical committees should. These reasons should be as cogent as possible, but in a pluralist society there should be no suggestion that the conclusions are those of an ethical oracle. They should be the result of reasoned examination of ethical problems by people representing no single strand of opinion in society. They should not claim to be more than this.

Finally, there is no need for the regulatory body to be large, or to have a large staff. A group of informed people thinking seriously about ethical problems is to be preferred to some huge and paralysing bureaucracy.

4. The international dimension

We envisage that regulatory bodies will operate at the national level; but we expect that members will note the decisions and reasoning of similar bodies in other countries. There is some hope for an international approach emerging, as a national approach may come from the interaction of local ethical committees.

In the case of research, it is particularly desirable that there should be as much international agreement as possible. A major problem in trying to see that science is used for human good rather than harm is that regulations stop at national frontiers.

Moreover, with issues like surrogacy, there is a special problem for the European Community. The free movement of labour makes it increasingly unrealistic to have quite different regulations about such matters in the different countries. If commercial agencies for surrogacy are banned in one country but not in another, this will hardly restrict their operation at all. Those who want to use them are unlikely to be deterred by a journey. Already some German couples wanting children make use of French sperm banks.

This makes a case for aiming at harmonization. Does this case take us further? Should there be a single European Community regulatory body, rather than a series of national ones? We think that this would not be a good idea at this stage. It is better to put up with a few anomalies than to attempt to impose consistency from the centre on matters touching on such deep values. Different religious and ethical traditions are strong in different countries. A European moral outlook will grow, if at all, out of the slow process of talking about things together. (And then, with luck, it will be more of a human moral outlook than a purely European one.)

Perhaps an agreed European set of policies will eventually emerge; but it will only do so by gradual evolution, through discussion of the reasons given for their decisions by the national regulatory bodies. When we look at a piece of reasoning, we often do so as human beings, rather than as Greeks,

Danes or Belgians. But the reverse is often true when some European regulation is imposed.

We should perhaps remember Kant's remark that out of the crooked timber of humanity no straight thing was ever made.

Part Five
Deciding who will be Born

Chapter 12
Handicap

Some uses of reproductive technology raise questions of special importance. Their answers can make a difference to who is born. To start with a misleadingly simple-sounding question: should we do what we can to ensure the birth of normal rather than handicapped babies?

1. The issues

(a) Antenatal tests and abortion: Antenatal tests make 'therapeutic' abortion possible. As well as the 'right to life' issue raised by any abortion, there is an additional question: does 'therapeutic' abortion discriminate in an objectionable way against the handicapped?

(b) Screening donors of semen or of eggs: If some donors are excluded because they are likely carriers of a severe disorder, this decision affects who is born. There is the thought: 'I might have had a different father,' or 'I might have had a different mother'. But this perhaps goes beyond the point where we have much grip on what of *me* there would be. (Suppose you are a blonde woman with two Danish parents, and the change of donor would have resulted in 'you' being a dark man with one Italian parent.)

(c) Screening of potential parents: If potential parents with serious psychological problems, or who are single or homosexual are denied access to reproductive help, this will prevent the birth of the children they would have had. In so far as such refusal of help is based on the interests of the potential children, the thought behind it is perhaps one that some kinds of environment constitute a form of 'social handicap', analogous to a medical one.
(d) Gene therapy: It may one day become possible to cure genetic disorders by direct intervention at the genetic level. We discuss the issues raised by this in the next chapter.

The dilemmas raised by these ways of affecting who is born are alarming partly because of a fear that we may be drifting towards 'eugenic' policies. But partly they are troubling in themselves. Who can feel comfortable about helping to bring into the world a child with a desperately severe handicap? But, on the other hand, who can feel comfortable about deciding whose life is or is not worth living?

2. Antenatal tests and abortion
The tests for some foetal abnormalities (Down's syndrome, spina bifida, etc.) make 'therapeutic' abortion possible. This term is sometimes used because there is a threat to the health of the mother, and sometimes because of the abnormality of the foetus. Opponents of abortion point out that this second use is euphemistic: in this case the 'therapy' eliminates the disorder by eliminating the patient.

There is the familiar question of whether abortion violates a foetal right to life, but with a new twist. Previously, the case for abortion centred on the claimed right to choose whether to have a child at all. The new dilemma is about aborting foetuses of a particular kind.

Among the questions raised are whether a severe handicap reduces or eliminates any right to life that a normal foetus would have; whether some handicaps are so severe that not to be born is in the child's own interests; whether other reasons, such as the effects on the family of a severely handicapped child, can justify abortion; and how severe the handicap has to be for any of these reasons to come into play.

3. Severe handicap: being born and not being born

Where children have very severe disorders, perhaps it would have been better if they had not been born. It is hard to be at all confident about this. People vary so much. Some people with terrible handicaps reveal extraordinary qualities, which enrich their own lives and those of people around them. So much depends on the individual case that generalizations are very uncertain.

One of the worst problems handicapped people have is the attitude of many other people to them. We are reluctant to say anything which might seem to downgrade the handicapped. We want to say that all people are of equal value, and that no-one is in the God-like position of being able to estimate the worth of another person's life. In a way these responses are absolutely right. Having a severe handicap is not a reason for being treated in any way as a second-class citizen. And, in the light of the way people have triumphed over handicaps, it is very hard to be sure that a particular handicap totally excludes a fulfilled life.

But, in one way, these decent responses may lead to a distortion of our thinking about these problems. Because we are rightly reluctant to say that a handicapped person must have a less fulfilled life, we may slide into saying that we have no grounds for preferring the birth of a normal child to that of a handicapped one. To have such a preference may start to seem like an objectionable form of discrimination against the handicapped. This line of thought makes it questionable whether we should follow medical policies such as genetic screening of donors, or therapeutic abortion, designed to reduce the incidence of handicap.

But these thoughts do not follow from the decent responses mentioned above. A particular handicap may not make a fulfilled life impossible, but it is not called a 'handicap' for nothing: it is likely to make such a life more difficult. Consider the theoretical possibility of deliberately causing a child to be born handicapped. This would surely be a monstrous thing to do. And we think this because we do *not* believe it is just as good to be born handicapped. Perhaps the best response is to retain a preference for the birth of a person without handicap, while rejecting any tendency for this view to spill over into discriminatory attitudes towards those who are handicapped. (It is possible that this spill-over cannot be avoided. If that is right, it is an important countervailing argument to set against the case for favouring the birth of an un-handicapped person.)

There are some legal cases in which it has been argued, on behalf of people with severe handicaps, that they have been harmed by being brought into the world with such handicaps. These 'wrongful life' cases are a fairly recent development. The idea first appeared in American lawsuits in which children born illegitimate claimed damages. In general, such claims were rejected. Later, 'wrongful life' suits centred on children born with severe handicaps or diseases.

Both parents and children acted as plaintiffs. The defendants were usually laboratories. Sometimes an action was brought against the parents themselves on the grounds of 'negligence' in not having opted for an abortion. The whole issue arises only in the context of the availability of antenatal tests and of therapeutic abortion.

The legal aspects of these cases are not our main concern. The central moral issue raised is whether parents or doctors ought to take what steps they can to prevent the birth of a child at risk of extreme handicap. To put it more strongly, does a child have a *right* to a life without severe handicap, or else a *right* to be prevented from being born?

One question about any such right is exactly who it imposes an obligation on. Do doctors or genetic counsellors have the duty to prevent the birth of such babies? This seems implausible. In general, their obligation is to provide information to enable the parents to take the decision. Perhaps things are different where the intervention of the doctors, through IVF etc., is required to bring the child into existence. But, if there is a right not to be born with severe handicap, it imposes duties mainly on the parents.

Then there are problems common to all claims that something is a right. In the first place, it may clash with other possible rights, such as that of those opposed to abortion to follow their consciences, or that of a woman bearing a child to take her own decision about the continuation of her pregnancy. Then, if there were such a right, there would be problems about its boundaries. We have seen that it is difficult to make general judgements about when a handicap rules out a fulfilled life.

These are good reasons for caution about accepting that there is a right not to be born with severe handicap. But it may still be true that in some cases it would have been better for the child not to be born. This is related to the questions of suicide and of voluntary euthanasia. Whatever is the best approach to those questions, they arise because some kinds of life are perhaps worse than not being alive at all. In one way, those questions are different, because the person is alive to make his or her judgement about them; but if it makes sense for people to see death as in their interests, there seems a parallel possibility of parents or doctors thinking that not being born may be in the interests of a potential child.

4. The comparison with non-existence

The claim we have considered is that it can be against the interests of a child to be brought into existence with an extremely severe handicap. This runs into a general problem of comparing existence with non-existence.

If, through these techniques, we determine that some people rather than others come into existence, can we say that those people are better or worse off for the intervention? There seems to be severe difficulties in making comparisons between being alive and any 'state' of being unconceived. These slippery conceptual problems make it hard to say that there is anyone who is worse off when a handicapped child is born rather than a normal one. And yet it does matter whether we conceive a normal or a handicapped child. We need a way of thinking about these problems which does not undermine this thought, yet which avoids sliding into paradox or absurdity.

One approach is to find some notional way of making the comparison with non-existence, perhaps by treating it on a par with being unconscious. Just as we prefer being unconscious to experiencing the pain of a major operation, so we could think that it is better not to be conscious at all rather than to experience certain kinds of life. Being dead or being unconceived could be treated as equivalent to permanent unconsciousness. This provides the kind of comparison we need, though it does so at the cost of a certain artificiality: there is a person who is unconscious for the operation, but there is no particular person in the state of being unconceived.

There is an alternative approach to this problem, which is worth considering in the context of less severe handicaps.

5. 'Moderately severe' medical disadvantages

We want to reject discriminatory attitudes towards people who are handicapped. At the same time we want to retain a view which most people would take as axiomatic: that it is better, where possible, to bring into the world a child without handicap.

This view, which we want to retain, runs into a problem which is often ignored. Take a handicap such as blindness. It is a substantial disadvantage to be blind, yet it is implausible to suppose that blind people might have been better off not being born. Such 'moderately severe' handicaps raise a problem for reproductive ethics. We want to stick to the axiom that we should, given the choice, bring into the world a child without such a handicap. But the means by which we may now hope to do this, such as donor screening or therapeutic abortion, act by preventing the birth of one child, with a view to substituting the birth of another.

The problem can be brought out by considering the position where we have not prevented the existence of someone with such a handicap. Has our failure to intervene done any harm? It is hard to say that the handicapped person has been harmed, for he or she is having a worthwhile life. (Such judgements are from the subjective point of view of that person.) There is no reason to think that the alternative (no life at all) would have been better for that person. Those of us who support, say, donor screening, have some difficulty in saying why it would be wrong not to screen for moderately severe handicap, given our inability to show that any particular child is harmed as a result.

One approach would be to base screening policies and therapeutic abortion on the interests of the parents in having a fully normal child. While the parental interests are important, there is much disagreement over whether they justify abortion. And supporters of therapeutic abortion may think that a case based entirely on parental interests does not capture all that they feel is at stake.

An alternative approach is to give up a plausible-sounding moral principle: that you do something wrong only when what you do causes someone to be worse off than they otherwise would have been. The principle has a platitudinous air, but reproductive ethics can make paradoxes out of apparent platitudes. Consider a reproductive thought experiment which casts doubt on the plausible-sounding principle.

There may be biological mechanism favouring the conception of normal babies over those that would be abnormal. Imagine a factory emitting a chemical which reverses one such mechanism. This mechanism now favours the conception of children with a moderately severe handicap. The pollution causes the existence of blind children rather than normal ones. The pollution seems to have done some harm. But to whom is the harm done? It has not made *those* children worse off than *they* would have been, since without it they would not have existed. The case that any particular person has been harmed by it is hard to sustain. Yet we are surely justified in seeing the pollution as harmful and in trying to stop it.

If this is right, reproductive ethics seems to be a field in which there are 'impersonal' harms and benefits. Harm can be done without there being identifiable people who are worse off than *they* otherwise would have been. In explaining why it is better to avert the conception of someone with a severe medical condition, we can use the idea of impersonal harms, without having to resort to metaphysical claims about benefits to a particular non-existent person.

The case for giving weight to these 'impersonal' harms can be supported by considering two cases closer to the central concerns of this report. The first case involves the 'womb leasing' version of surrogate motherhood. If there were a choice between different women wanting to act as surrogate mother, where other things are equal, there is a case for choosing a non-smoker rather than a smoker. The same child would be born, but without any damage that smoking would have caused. The second case involves the other version of surrogate motherhood, where the woman donates an egg. Where other things are equal, should a non-smoker be preferred? Some may think, as we do, that the choice in the second case should go the same way as in the first. But, as the child in the second case can reasonably be thought to be a different person where the egg is different, those who think this have to cite some 'impersonal' reason in explaining why they do so.

6. Where the alternative is not a different child, but no child

In the two surrogacy cases just considered, none of the choices excluded *some* child being born. But what if it is a choice between the risk, or even certainty, of some disadvantage, and deciding against having a child at all?

Where the disadvantage is so appalling that life can reasonably be thought to be not worth living (the kind of disorder that gives rise to 'wrongful life' suits), there is little problem. It is obviously wrong deliberately to bring into existence someone whose life you know will be terrible.

There is more of a problem where the condition comes into the category of what we called 'moderately severe' disadvantage. The handicap might be something like blindness or cleft palate. If the choice is between giving reproductive help to produce such a child, and refusing with the result that no child is born to that couple, what is the right thing to do?

Here we report a conflict between a widely held feeling, which we have too, and the results of looking in a more detached way at the two outcomes. The widely held feeling is that it surely cannot be right deliberately to produce a child you know will suffer from a fairly severe handicap. But on the other side are these considerations: the couple want the child, and the child, while regretting the handicap, will not regret having been born. So no-one is made worse off by the decision to have the child. And, as there is no alternative of having another child without the handicap, there is no 'impersonal' harm. This committee has no consensus view on this question. Some of us think it is right to give reproductive help. But we find the issue difficult, and report the conflict between intuition and consequentialist analysis for the reader of this report to consider.

Chapter 13
Gene Therapy, Genetic Engineering and Sex Selection

It has been estimated that at least ten per cent of diseases are genetic in origin. There is at present no cure for genetic defects. There are two major possible developments. One, already with us for some disorders, is to screen for certain genetic defects and, using IVF, to transfer only healthy embryos. The other possibility, more speculative and problematic, is that we shall be able to use 'gene therapy': the replacement of defective genes by normal ones.

We should bear in mind also the related possibility of genetic screening shortly after birth followed by preventive treatment, at present applicable to only a few disorders (for example phenylketonuria, of which the effects may be minimized by controlled diet). Such post-natal screening may in future have far-reaching effects on genetic diseases.

1. Screening for genetic disease

The use of IVF, together with screening pre-embryos before replacement, in many cases makes it very probable the mother will have a child free from a genetic disorder known to run in the family, without having to have an abortion.

Advances in molecular biology have given new tools for prenatal diagnosis. So far, these have mainly been used for early detection of embryonic or foetal abnormalities, quite independently of reproductive technologies. But applying these techniques to cells from an implanted embryo (using amniotic fluid or chorionic villus sampling) is not different in principle from applying them to an embryo not yet implanted.

Some genetic disorders can be directly or indirectly tested for, using recombinant DNA techniques.

The direct test (used, for instance for sickle cell anaemia) applies a specific cutting enzyme to the DNA to be tested. The cutting enzyme is directed to a binding site which marks the presence or absence of the defect. An abnormal cutting fragment indicates the presence of the defect.

Where the location of the defect is not yet precise, or where a specific cutting enzyme is not available, indirect tests are sometimes possible. (Such cases include cystic fibrosis, Duchenne's muscular dystrophy, Huntington's chorea, and familial Alzheimer's disease.) One method (whose accuracy is still variable) exploits links between abnormal genes and nearby repetitive DNA patterns. It is possible to clone DNA markers, which bind to the DNA pattern linked to the defect.

The size of the DNA sample needed for these prenatal diagnoses has meant that cells (whether from the amniotic fluid, from chorionic villus tissue, or from an unimplanted pre-embryo) have needed to be cultured for several days. But this can now be avoided by using polymerase techniques which enable the DNA to be copied 200,000-fold in several hours.

Where neither direct nor indirect recombinant DNA techniques are available, genetic defects can often be detected by biochemical indicators, such as enzyme levels in cultured cells. In this way, such abnormalities as Lesch-Nyhan syndrome, Hunter's disease, Hurler's syndrome, and Tay-Sachs disease can now be detected.

It is worth remembering what these diagnostic techniques can mean to people. Consider first cystic fibrosis, now detectable by indirect recombinant DNA techniques. It is caused by a recessive gene carried by one in twenty Europeans, so that one in every four hundred couples risks having an affected child. Each of their children has a twenty-five per cent risk of the disorder.

A couple who have already aborted two foetuses with cystic fibrosis may not want to risk conceiving again. IVF and screening could be used for couples at risk, giving them the chance of a normal child without having to abort.

Or, to take another example, the (rare) Lesch-Nyhan syndrome is among those now detectable on the basis of enzyme levels in cultured cells. This is a single gene defect transmitted on the sex chromosome. The defective gene results in the enzyme hypoxanthine-guanine phosphoribosyl transferase (HGPRT) not being produced. The children seem healthy when born, but within weeks or months they develop cerebral palsy. They are severely mentally handicapped. They make constant writhing movements, and the build-up of purines in their bodies brings continual pain. The children involuntarily mutilate themselves, often biting off parts of their tongue and lips. Sometimes, the only way to prevent this is to take out all their teeth. These children do not die until they are teenagers.

It is not hard to understand why people who want to be parents, but whose potential children are at high risk of such a disorder, may see genetic screening as a mercy.

2. Gene therapy

With disorders caused either by the absence of a gene or by the presence of a 'wrong' gene, it is attractive to think of inserting or deleting genes as required. Perhaps the day will come when we can do this. But first there are serious ethical and practical problems to consider.

It is important to distinguish between germline and somatic gene therapy. Germline gene therapy involves the insertion of genes into fertilized eggs or into very early pre-embryos, so that the new genes enter germ cells or cells which give rise to them. Somatic gene therapy changes cells already differentiated, for instance blood cells. Germline therapy leads to the new gene being passed on to subsequent generations, while somatic gene therapy does not.

Germline gene therapy involves serious risks. These are illustrated by some work on inserting genes into mice, carried out in connection with cancer research. Some drugs used to treat cancer switch off folic acid metabolism. The aim was to produce mice resistant to these drugs by inserting a gene that itself switches off folic acid metabolism. The gene was sequenced and injected into mouse embryos. Of those embryos which survived and were implanted, significant numbers had incorporated the gene into the germ cell line, so passing it on to their descendants. Most of the mice incorporating the new gene had visible abnormalities, which suggests that other genes may have been displaced by the new gene.

Because of the risks of similar problems, there is a very strong practical case, in our current state of knowledge, for rejecting the application of this kind of gene therapy to humans. This is quite apart from ethical concern about changing the genes of the person's descendants, which has resulted in a policy being adopted by European Medical Research Councils that germline therapy in humans 'should not be contemplated' ('Gene Therapy in Man', a joint statement by the Medical Research Councils of Austria, Denmark, Finland, France, The Netherlands, Norway, Spain, Sweden, Switzerland, the United Kingdom, and West Germany, *Lancet*, 1988).

Somatic gene therapy does not differ in principle from organ transplantation, and does not raise ethical issues about future generations. The field is still at an early stage of development, and so far applications to animals have not resulted in any successful cures. Its use in humans depends on obtaining more promising results in animal studies, as well as on having grounds for confidence that the risk of harmful side-effects (such as the activation of genes involved in cancer) is very low. The requirements of low risk and a reasonable chance of benefit are standard ones for medical intervention. Somatic gene therapy does not seem to us to raise special problems of principle which would require any further restrictions. But it is worth noting that, where a genetic defect is identified in a pre-embryo, it will often be possible to screen other pre-embryos with the same biological parents. Finding an unaffected one to transfer makes gene therapy unnecessary.

3. Positive genetic engineering

Safe gene therapy is not yet with us. But it is better to start thinking about the implications of likely technological developments before they arrive. The stage when a technology seems too remote to be worth thinking about has been followed so often by abrupt transition to the stage when it is already out of control. If this 'negative' gene therapy does come, one worry is that it may be part of a slide towards other, non-medical interventions of a 'positive' kind, designed to make 'improvements' in people. The technology for this is even further off. But the eventual mapping and sequencing of the human genome, together with knowledge of how genes affect characteristics, may one day make positive genetic engineering possible.

One view is that, just as a child can benefit from an upbringing which makes him or her more intelligent, more musical, or better at making friends, so a child could equally benefit from genetic changes which are found to have the same effects.

The other view is that it is in the interests of the child not to be genetically the product of other people's decisions. As the European Parliament put it (Recommendation 934, 1982), there is a 'right to genetic inheritance which has not been artificially interfered with, except for therapeutic purposes'.

Positive genetic engineering is in many ways an extremely disturbing prospect. But it is debatable whether the issue is best seen in terms of violating the rights of those whose genes would be altered. We have seen the unsolved problems in marking off rights from other interests. And a particular problem about this right is how far people would in fact feel that their interests had been overridden. Some people might greatly dislike the fact that their genes had been chosen for them; but others might feel that they had gained from having genes making them more intelligent or more amiable, and be glad that their genes were not left to the natural lottery.

It might be hard to convince people in this second category that a right of theirs had been violated. Perhaps the fact that some might be glad of the intervention is not enough to justify it. But there are complex issues here, and there is a suspicion that the 'right' has been plucked out of the air to settle a difficult issue at great speed.

There are real causes for concern about positive genetic engineering, and it is worth bringing them into sharper focus.

One concern is perhaps obvious to any late twentieth-century European. This is the danger of giving new technological powers to a government with a programme of racist eugenics, or perhaps to one which had some other, not 'race'-linked, category of people it regarded as less than fully human. In our countries there is nearly-universal revulsion against this kind of outlook. But technology once developed is always available, and it is hard to be sure that such an outlook will never return.

Another cause for concern is not the deliberate use of genetic engineering for perverted ends, but the accidental side-effects of well-intentioned uses. One problem is that we may overlook some genetic linkage between different characteristics. In choosing, say, genes for high intelligence, we may find we have inadvertently also chosen genes for high levels of aggression or depression. All technology carries with it the risk of mistakes and unwanted side-effects. Here the 'mistakes' will be human beings. There is also the risk of disasters which are irreversible: 'mistaken' genes would be passed on to future generations.

Another problem is that if certain characteristics were frequently selected, those who did not have them might come to be seen as inferior. It may be hard to keep the decent reluctance to see people in terms of some competitive rank ordering, when, in another part of our mind, we are ranking characteristics which may be available for our children.

There are also enormous problems about who would take the decisions. These problems are sometimes gestured at by describing the choice as 'playing God', with the implication that it would take a great deal of presumption to put oneself forward as qualified to choose the genes of another person.

Perhaps, if there were great benefits, we would overcome our reluctance to play God in this way. But there would remain the question of whose decision it should be. Many shudder at the idea of some central government committee taking such decisions. The alternative of parents choosing for their children is much more attractive. But that too has huge unsolved problems. Could parents not sometimes choose disastrously for their children? And suppose too many parents chose the same characteristics? What social limits to their choice would be set, and by what means?

To some, these dangers and problems are enough to support an absolute rejection of positive genetic engineering in principle. To others, these are very serious objections, but not enough to justify the certainty that it could in no circumstances be acceptable. On this second view, the possibility of benefits great enough to outweigh the risks and drawbacks cannot be ruled out.

Even on the more permissive view, the risks and problems are still very great. We are nowhere near having thought through the problems of how the decisions should be taken, nor those of what the social impact of different applications would be. It is just as well the technology is not with us yet. And if it does start to come over the horizon, we should simultaneously do two things.

The first is to anticipate, in the light of the technical knowledge available, what kinds of application will become possible. Then it will be necessary to think which applications, if any, will bring benefits worth having. Then the right mix of individual and social participation in the decisions will have to be worked out.

The second, since the problems are so great, and the consequences of the wrong decisions are so bad (and probably so permanent), is to accept the need for a breathing space, so that we stand a better chance of getting the decisions right. This gives a reason for at least a temporary ban on non-medical genetic intervention. (We accept that the line between medical and non-medical intervention is often a blurred one, and that any line we draw will, like the speed limit, have an element of arbitrariness.) Those who oppose such non-medical intervention on principle will favour a permanent ban, but even those who are more optimistic about its possibilities should accept that the problems are important enough for it to be very foolish to start using the technology before we have come up with some solutions.

This adds up to a case for saying that positive human genetic engineering is unethical now (or when it becomes available) and will remain so at least until policies have been worked out to cope with the huge problems it raises.

4. Sex selection

Many parents are happy to leave the sex of their children to nature. But some potential parents have a strong preference for one sex. This can be because of an intrinsic preference for children of a particular sex, perhaps a 'traditional' preference for males or a 'feminist' preference for females; or it can arise from the desire in parents who already have children of one sex to have both boys and girls in the family.

Those who have such strong preferences may desire an abortion where tests show a child of the other sex. The development of chorionic villus biopsy makes abortion on the basis of sex possible before the twelfth gestational week, and so within the time-limit placed by many legal systems on abortions.

Less traumatically, those with such preferences may seek to use methods of pre-conceptual sex selection. One such method, recently developed in Japan, consists in the centrifugation of sperm cells. This was reported to be ninety per cent sucessful in separating X from Y sperm, due to their different density (*Nature*, 321 (1986): 720).

(a) Arguments in support of sex selection: Sometimes sex selection has a medical justification. Where parents know that they have a high risk of producing a child with a sex-linked disorder, their case for wanting a child of the other sex is quite unproblematic. The difficult issues begin when parents want to choose the sex of their child for non-medical reasons.

It can be claimed that these techniques are an extension of

freedom. In ordinary life, there are social pressures which affect choices about having children. For instance, a thirteen-year-old girl who is pregnant may be subject to strong pressure to have an abortion. But, in general, choices about whether to have children at all, and about how many to have, with all their implications for those who may be born and for society, are left largely up to parents. Given this, it may seem arbitrary to rule out choice of sex, especially if this matters a lot to the parents. Already, whenever we implant a particular embryo, we have in effect chosen the sex of a child. If we can determine sex as a side effect, it may seem unreasonable to exclude doing so by choice.

(b) Some comments on these arguments: We have no reservations about sex selection in the special case where it is undertaken to avoid a sex-linked disorder. Our doubts start with the arguments for choosing sex in other cases.

It is true that the use of these techniques would extend parental choice, and would enable the satisfaction of preferences which might otherwise be frustrated. But these points have to be weighed against the possible harmful consequences of sex selection. And there are grounds for some fairly strong reservations about the preferences it would satisfy.

One alarming possible consequence is a serious sex imbalance in the next generation. Perhaps those who will be members of that generation have an interest in averting this. And, in the nature of things, their voice will not be heard at the time the decisions are taken.

This raises the question of how we decide what ratio is unacceptable. Is the ideal an exact balance between males and females, or some departure from equal numbers? Or does the ratio not matter at all, or not matter much? And how likely is sex selection to lead to an unacceptable ratio? We do not claim to have definitive answers to these questions. But it seems reasonable to have a strong presumption in favour of a rough equality of numbers. This is not only the traditional ratio, but it also seems likely to make everyone's sex life easier. Any policies leading to a large imbalance would need some strong justification.

One justification offered is the satisfaction of parental preferences. But there are grounds for viewing these preferences without enthusiasm. Of course, people should not be prevented from doing things just because others do not approve of their preferences. But the defects of the preferences influencing sex selection may harm others.

In the first place, the attitude to the child is one which it may be better not to encourage. It is unconvincing to suggest that the child will benefit by being one gender rather than another. This is not just the identity problem again: the question of whether it will be the same child. It is a simple point: there is no known general truth that girls are happier than boys, or vice versa. The preference is surely not for the sake of the child, but for the parents themselves. Many parents put the interests of their children first, and this other, more egocentric, attitude is one which may be less beneficial to the children.

Another problem is that technology does not always work. If a child of the 'wrong' sex is born, such a strong preference for the other sex may conflict with the unconditional acceptance which is important for children to flourish. There is something more warming about a parent who agrees with Betty Hoskins and Helen Holmes: 'A reasonable stance in the case of sex preselection would be *not* to choose a girl or a boy, a boy or a girl, but to welcome each *child*' ('Technology and Pre-natal Femicide', in Rita Arditti *et al*: *Test-tube Women*, London, 1984).

There is also the possibility that the choice will sometimes be an expression of sex prejudice, reflecting an attitude that one sex is in general inferior. It is better to undermine the attitude than to satisfy the demand it creates.

And, even where sex prejudice is absent, there may be other grounds for concern about the attitudes displayed. One American study in the 1970's of married women under forty found that twice as many would prefer a boy to a girl, (L.W. Hoffman: 'Social Change, the Family and Sex Differences', paper given at National Council for Family Relations, 1976). The three main reasons they gave were to please their husbands, to carry on the family name, and to provide a companion for their husband. There is a generosity towards their husbands in these attitudes. We hope the men had an equally frequent preference for girls, based on equivalent thoughts about pleasing their wives. But we can imagine a world without this symmetry, and in *that* world the inequality of outlook would cause concern.

And, if one sex is chosen much more, this may make it harder to have a climate of equal respect between the sexes. Whatever people *say* about it being just as good to be either male or female, can we expect a girl to believe this in a society where most people, perhaps including her own parents, have chosen to have two boys and one girl?

5. A suggested approach to sex selection

Leaving aside the case of sex-linked disorders, it could be argued that there is nothing *intrinsically* wrong with sex selection. Suppose eighty per cent of people were already doing it, so that the 'natural' sex ratio had already disappeared. It is hard to see what specific objection could be made if the remaining parents wanted to use sex selection to swing the ratio back towards equality. And the same point could be made in the smaller context of a family. There is nothing objectionable about the desire for a child of one sex to counterbalance the two of the other sex you have already had.

But we are impressed by the dangers of an unbalanced sex ratio, and inclined to the view that the desires behind the choice of sex will often be ones society would do better to discourage than to satisfy. So we think that sex selection should be strongly discouraged. It would be desirable for clinics to be banned from providing pre-conceptual techniques as a service. This would reduce the incidence of sex selection. It would not exclude abortion on grounds of sex: to try to ban this would be intrusive, and would probably be ineffective. But we think that the numbers prepared to go to the lengths of having an abortion would be relatively small.

We think it desirable to limit the intrusion of the law into individual decisions about these matters. It is more acceptable to control institutions such as clinics than for the law to attempt directly to restrict the actions of the parents. So we accept that, if safe 'do-it-yourself' methods are developed, these should not be banned. Perhaps, in the long term, our best hope is the erosion of the attitudes which make sex choice seem so important.

Part Six
Summary of Conclusions

Summary of Conclusions

Since this report deals with ethics at least as much as with public policy, there is something dogmatic about a bald list of assertions to the effect that this is right and that is wrong. Before listing our conclusions, we wish to stress that their force can only be assessed in the context of the argument which has led to them. A report of this kind makes more of a contribution when people are encouraged to think more deeply, and probably to disagree, than when they uncritically accept the conclusions.*

1. Parents, donors and children

1. We think there should be a *presumption* that those born as a

*Dr. Simone Novaes, while not dissenting from the report as a whole, has two reservations:

1. The report does not sufficiently discuss the role of physicians and other medical intermediaries in the way these new reproductive technologies are used. Because of the role of the medical profession, there is a tendency to see these technologies in terms of a medical model: as ways of 'treating' infertility. However, the medical profession is making available, for procreative purposes, a medical alternative to sexual relations. In AID this is particularly clear. This is what,

result of semen donation should ultimately have access to knowledge of the identity of their biological father, but that this presumption could be overridden by evidence that it drastically reduced the supply of donors.

There is a case for an *experiment* on Swedish lines, where for a trial period:

(a) the child is given a legal right to know the father's identity on reaching maturity.

and

(b) legal paternity is assigned to the married woman's husband, and donors are given protection against any legal claims.

A severe and continuing shortage of donors would be a good reason for reversing the policy after the trial period.

2. We think that there should be no general presumption for or against related donors. But, in the interests of donation being fully voluntary, daughters should not normally donate eggs to their mothers.

3. In the case of related donors, the child should not be deceived about the relationship.

4. We are divided on whether reproductive technology should be made available to people other than infertile heterosexual couples. But we agree that the birth of a child should not be associated with criminality; and consequently we agree that no

---

in the long run, makes reproductive technology such an explosive issue. It automatically transforms reproduction into something more than an intimate family matter. The role of the medical intermediary in this transformation deserves fuller treatment.

2. The list of conclusions may induce a false sense of security, at least among those inclined to agree with them. They seem too reductionist to do justice to the complex arguments of the report. It would have been preferable to have had conclusions along more general lines, summarizing the different groups of problems, as well as new questions and lines of thought, with less suggestion that we might have solved some of them. This might have been more helpful in stimulating reflection.

use of these techniques by individuals or couples should be illegal.

2. Surrogacy

1. We favour a restrictive approach to surrogacy, in the interests of protecting the surrogate mother from exploitation, and in the interests of protecting the child from harmfully prolonged battles between the surrogate and the prospective parents.
2. Commercial agencies should not be permitted.
3. Any public agencies set up should help people to find a surrogate mother only when there is a clear medical reason for doing so.
4. To avoid criminalizing the birth of a child, private surrogacy arrangements between individuals should not be illegal.
5. If a surrogate makes a contract, it should in certain respects not be enforceable: whether or not she has an abortion, and whether or not she hands over the child, should not be matters over which she is legally compelled.
6. If the surrogate does hand over the child, she should have no right to claim the child back.
7. Except where the surrogate is in the same family, the normal practice should be for relations between her and the child to be severed.
8. Agencies should screen potential surrogate mothers, on medical and on psychological grounds.
9. We are not enthusiastic about payment to the surrogate mother, beyond the meeting of her expenses. But we are agreed that such payment should not be illegal.

3. Research

1. We value the benefits which research on pre-embryos may bring.
2. Despite having differing views on the moral claims of the pre-embryo and embryo, we agree that they are not strong enough to exclude all research, or even to exclude all research which destroys the life of an embryo.

3.We agree that research is easier to justify at early stages of development than at later stages.
4. The stage of development at which research should no longer be permitted should be decided by a regulatory authority with powers to vary the boundary in the light of new evidence.
5. No research causing foetal pain should be carried out.
6. Where an embryo or pre-embryo will eventually become a child, to avoid the risk of harm to that child no research on the pre-embryo or embryo should be carried out.
7. We see no objection to the use of material from aborted foetuses for medical purposes such as transplants. But such material should not be used for non-medical purposes.
8. Transplants should be obtained from foetuses aborted on other grounds, and the beneficiary should normally be unknown to the woman having the abortion.
9. We see no objection to cultivating cells of a particular kind (e.g. blood cells) for medical purposes, nor to this cultivation going beyond the time-limit set for research on the whole embryo.
10. Research should be regulated by a statutory body, to which ethics committees would refer any decisions involving major new issues of principle. This body should be independent of government, and among those well represented on it should be women and those who are neither doctors nor scientists. It should make public the reasoning behind its decisions.

We recommend that, if the scope of the activities of the body is broader than research, and includes the whole field of new reproductive technology, it should commission a large-scale study of the children resulting from these kinds of intervention. This should include investigation of the psychological effects on children of having a donor in the family as against the donor being unknown.

At many points, we have been struck by the high ratio of assertion to evidence on these matters. A substantial study would enable people a generation from now to make far better based decisions than ours.

(An unavoidable dilemma is that often we cannot be sure

what the right policy is without evidence, but the evidence only comes from trying out policies. The best response is to go carefully, choosing what looks the best policy, but following up the consequences and being prepared to change as a result. The same dilemma is familiar to doctors trying out treatments for illness. Doctors sometimes say that the *only* ethical form of medicine is experimental. This may be true of social policies too.)

11. Each country should have its own regulatory body. There should not be an imposed European policy.

4. Deciding who is born

1. It is desirable that donors should be screened to reduce the risk of the birth of a severely handicapped child.
2. We accept that abortion in cases of serious handicap is justifiable.
3. The objection to gene therapy of somatic cells is its high level of risk. If this could be overcome, we see no objection to its use.
4. If 'positive' genetic engineering becomes possible, it should not be permitted until it becomes clear (if it ever does) that the problems it raises can be overcome.
5. Clinics should not be permitted to offer sex selection of children, except on medical grounds.
6. 'Do it yourself' sex selection should not be illegal.

Bibliography

Some reports on these issues.

Barn Genom Insemination (Children Conceived by Artificial Insemination), Justitiedepartement, Stockholm, 1983.

Human Procreation, Ethical Aspects of the New Techniques, Council for Science and Society, Oxford, 1984.

Report of the Committee of Enquiry into Human Fertilisation and Embryology, London, 1984 ('The Warnock Report').

Committee to Consider the Social, Ethical and Legal Issues arising from *In Vitro* Fertilization: *Report on the Dispositions of Embryos Produced by* In Vitro *Fertilization*, Victoria, 1984.

Ontario Law Reform Commission: *Report on Human Artificial Reproduction and Related Matters*, Toronto, 1985.

Report of the Commission on Human Artificial Fertilization, Rome, 1985 ('The Santosuosso Report').

Report of the Commission on In Vitro *Fertilization, Genome Analysis and Gene Therapy*, Bonn, 1985 ('The Benda Report').

Ethics Committee of the American Fertility Society: *Ethical Considerations of the New Reproductive Technologies*, Birmingham, 1986.

Congregation for the Doctrine of the Faith: *Instruction on Respect for Human Life in its Origin and on the Dignity of Procreation*, Rome, 1987.

Étude du Conseil d'État: *Sciences de la Vie, De l'éthique au droit*, Paris, 1988.

Index